Jahja Kokaj

Biomedical Deformation Measurement using Laser Techniques

AF141044

Jahja Kokaj

Biomedical Deformation Measurement using Laser Techniques

LAP LAMBERT Academic Publishing

Impressum / Imprint

Bibliografische Information der Deutschen Nationalbibliothek: Die Deutsche Nationalbibliothek verzeichnet diese Publikation in der Deutschen Nationalbibliografie; detaillierte bibliografische Daten sind im Internet über http://dnb.d-nb.de abrufbar.
Alle in diesem Buch genannten Marken und Produktnamen unterliegen warenzeichen-, marken- oder patentrechtlichem Schutz bzw. sind Warenzeichen oder eingetragene Warenzeichen der jeweiligen Inhaber. Die Wiedergabe von Marken, Produktnamen, Gebrauchsnamen, Handelsnamen, Warenbezeichnungen u.s.w. in diesem Werk berechtigt auch ohne besondere Kennzeichnung nicht zu der Annahme, dass solche Namen im Sinne der Warenzeichen- und Markenschutzgesetzgebung als frei zu betrachten wären und daher von jedermann benutzt werden dürften.

Bibliographic information published by the Deutsche Nationalbibliothek: The Deutsche Nationalbibliothek lists this publication in the Deutsche Nationalbibliografie; detailed bibliographic data are available in the Internet at http://dnb.d-nb.de.
Any brand names and product names mentioned in this book are subject to trademark, brand or patent protection and are trademarks or registered trademarks of their respective holders. The use of brand names, product names, common names, trade names, product descriptions etc. even without a particular marking in this works is in no way to be construed to mean that such names may be regarded as unrestricted in respect of trademark and brand protection legislation and could thus be used by anyone.

Coverbild / Cover image: www.ingimage.com

Verlag / Publisher:
LAP LAMBERT Academic Publishing
ist ein Imprint der / is a trademark of
OmniScriptum GmbH & Co. KG
Heinrich-Böcking-Str. 6-8, 66121 Saarbrücken, Deutschland / Germany
Email: info@lap-publishing.com

Herstellung: siehe letzte Seite /
Printed at: see last page
ISBN: 978-3-659-61399-9

BIOMEDICAL DEFORMATION MEASUREMENT USING LASER TECHNIQUES

Jahja Kokaj

SCIENTIFIC CENTER
COSMOS AND HUMAN
Gjakova, Kosova 381390
Email: jkokaj@yahoo.com

1

Acknowledgement:

I thank Prof. Dr. Goran Pichler, academician of Academy of Science of Croatia, for reading the final version of the manuscript and for editing the book.

Mr. Eng. Joseph Mathew for the technical assistance required for the book.

Table of Contents

Introduction

Laser and fiber optics applications have opened some new possibilities in medicine and dentistry. Advanced diagnostic techniques and laser-based therapeutic techniques have been developed.

Here, we have introduced some new techniques for medical and dental applications using lasers and modern optics. The main task of our work is biomedical applications of lasers.

In the first chapter a new approach based on image holography is applied for dental deformation measurement. The advantages of this technique compare to existing techniques are shown. Imaging and deformation measurement of in-plane and out-plane deformations is performed using the interferometric fringes. Interferograms obtained by holographic- based techniques are shown to

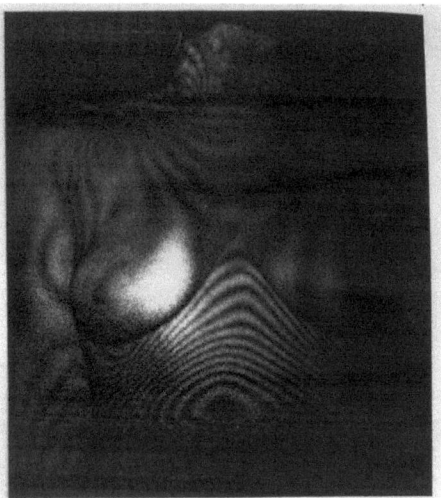

Double exposure holographic interferogram[from S. M. Zivi and G. H. Humberstone " Chest Motion Visualized by |Holographic interferometry" Medical Research Eng. P. 5 (June 1970)*

4

be useful to perform contour analysis leading to 3-D information of an biomedical object. Such an example for chest investigation using holographic interefometry is shown in the figure above performed by Zivi and Humberstone*.
The fringes are projected on the reconstructed holographic image of the dental components to be investigated.

In second chapter we have shown a new laser technique for biomedical application. This is based on moments and Fourier analysis. This imaging technique, based on optical implementation of moments, as to our best knowledge, is introduced for the first time in biomedical applications.
In addition to this chapter Fourier Transform of far Field diffraction technique is introduced as well.

Since the application of moments is a new technique for deformation measurement, in the third chapter, some details are elaborated. In order to increase the sensitivity and accuracy of this technique, higher order of moments are introduced and analyzed. We have shown that by using higher –order moments the accuracy of the measurement is increased. We have shown the advantages of the use of higher diffraction orders as well. However, we have shown the experimental limitations when these techniques are used. The accuracy measurement depend on the wavelength of the coherent light or laser light.

In the fourth chapter a simple and robust technique based on the Moiré phenomenon is applied.
Although the mathematical formulation of moiré phenomenon is introduced Moiré than one century ago, here is shown to be useful in application in modern technology. The mathematical or theoretical analysis and presentation of moiré phenomenon is out of the scope of this book. However we have borrowed some written material from different sources and projects such as NASA project and placed at the end of this book as an appendix. We would like to emphasize that appendix is not our intellectual property but is browed from different sources and outdoors, and proper citations are provided.
Using Moire fringes generated by two gratings and their projection on the object to be analyzed, contours and 3-D information can be obtained. We have studied biomedical simulated deformation of the human back. Detailed analysis and fringe interpretation is shown. Fringe interpretation and quantification deformation performed in this chapter could be applied in the previous Sections as well.

Section 1.

Dental deformation measurement using coherent optical techniques

Deformations of the dental model are visualized by applying image holography and Abramson's method. In-plane, rotation and out-of-plane deformation are detected. For the measurement of the deformations a diffraction grating method is introduced which is shown to have high sensitivity, accuracy and ability to detect 3-D deformations.

1 Introduction

A wide range of optical methods are available for deformation measurements including holographic interferometry [1-7], Moire technique [8-10], Deformation Measurement using Moments (DMM) [12], and other coherent techniques [13-20].

For deformation measurement of human bones, dental parts, and artificially made limbs sometimes is preferred to use two or more complementary optical or other Non-Destructive Methods (NDM).

In Section, holographic interferometry and diffraction grating method are applied as complementary methods. Holographic method is discussed in section 2. A pressure is applied to indicate, detect and recognize the types and segments (regions) of the deformations. The obtained results and extracted regions of the object, from the reconstructed imagery are used to perform the deformation measurement. On the extracted regions a new introduced method is applied. We call it Diffraction Grating Method (DGM). This is presented in section 3. The sensitivity and accuracy of the measurement is discussed in section 4. The conclusion is presented in section 5.

2 Application of image holography interferometry

A mechanical device is used to introduce and measure the pressure on a dental

model. The convex side of the model is shown in the fig. 1a and its concave side is shown in fig. 1b.

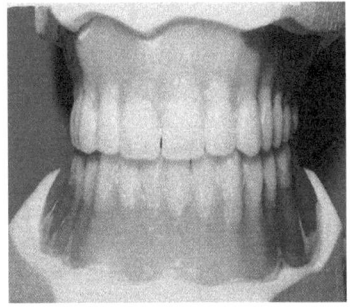 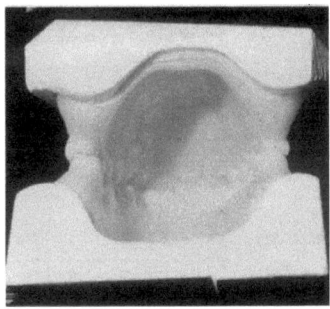

Fig. 1.1 Images of convex and concave view of a biomedical object

The real conditions and deformations were simulated like the model were used in real activity. In-plane scaling, rotation and 3-D deformation will be exercised during the simulation of the chewing-eating process. In case the chewing is not symmetric and is not acting simultaneously in both sides, the rotation or twisting deformation will occur. Therefore on the model under pressure, different types of deformations in the different regions will be produced. They are visualized by the interferometric patterns which are superposed on the reconstructed image.

The experiment is performed on a vibration controlled table. The collimated laser beam illuminates the concave side of the object. A diffracted beam, so called object beam passing through lens has projected the image of the object on a holographic plate. The hologram of the object is obtained by illuminating the plate by a reference beam. The figure of the setup and detailed description here are omitted since they are shown elsewhere (this is ordinary procedure of image holography) [1-6]. The next hologram is obtained after a force of 73 N introduced on the model.

During the reconstruction both holograms were placed at the same position. They were illuminated with the reference beam so the reconstructions of two wavefronts were obtained. Due to coherence of the reconstructed beams an interferometric pattern is produced. This can be seen on the surface of the reconstructed imagery. The pictures of the different parts of the same reconstructed imagery has been taken by changing the angle of view of the camera. A telecentric system was applied when the pictures were taken. The obtained pictures of the certain regions are shown in the figs. 2a, 2b and 2c. These figures has been taken from the same imagery only by changing the angles of the view of the camera of each exposure respectively.

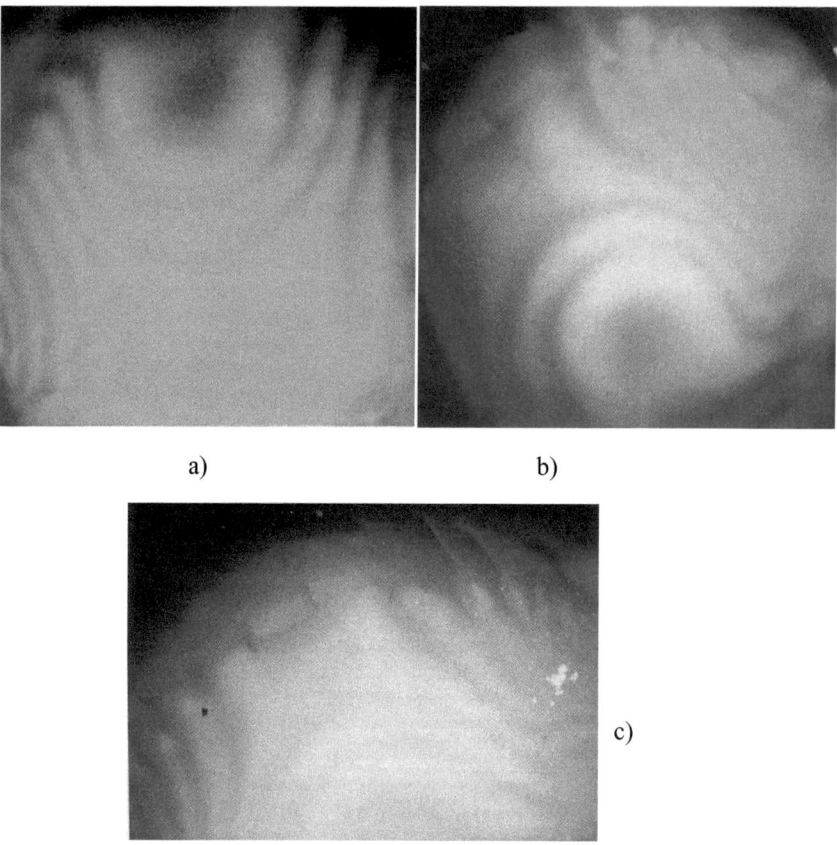

a)   b)

c)

Fig 2. Images of reconstructed holograms obtained by 3 different angles of views.

The fringe interpretation, region extraction and identification of the deformation types has been performed. We will not elaborate this in this paper since it is a routine procedure [1-6]. However it is shown that in-plane, rotation and 3-D deformation appears in the model when the pressure is introduce.

In the following section, the technique we proposed for the deformation measurement is shown.

3. Application of the diffraction grating for the deformation measurement

A pattern of a grating or a grill is projected on the regions of the concave side of the model surface (shown in fig. 1b). A transparency is made by photographically copying projected pattern on the model surface. The obtained transparency, which actually is a grating, is placed on the plane P1, shown in the fig. 3. By illuminating the grating with collimated laser light the Fraunhoffer diffraction or Fourier transform is obtained.

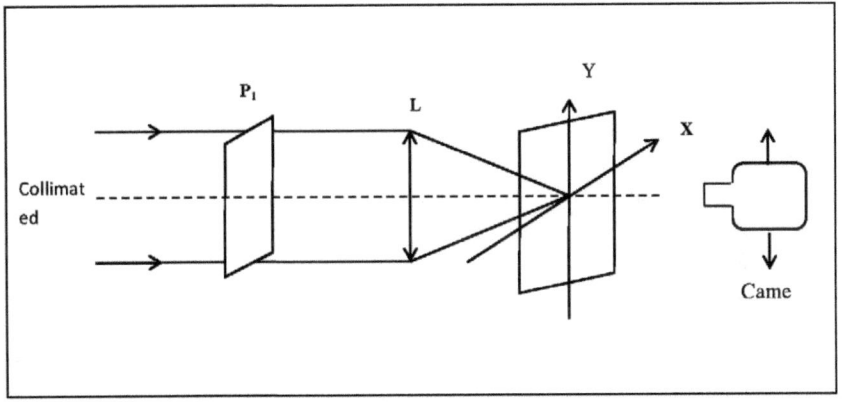

Fig. 3. Coherent optical Fourier transform system

The changes based on the deformation of the object will change the constant (the pitch) of the grating. Therefore 2-D position and inter distance of the diffraction spots on the screen tell us about the deformation of the object under investigation.

Fourier transform of a function gives much information about the function itself. By analyzing changes in Fourier transform, one can determine to some extent what has happened to the function. If the grating is printed on the model surface, the model deformation will destroy the grating. The changes in Fourier transform of the destroyed grating enable one to determine the deformation of the model changes in grating pitch and changes in orientation of Fourier transform of the grating enable to determine the complete deformation parameters; the strain components in two orthogonal directions, say x and y, plus the corresponding angle deviation, or the strain component in that direction. The model should be provided with a grille (crossed grating) to fulfill those criteria. For convenience, we will carry on the analysis for the grating, the results can be readily extended to the situation of a grille. A space-line grating can be written as a 1-D function.

$$f(x) = \mathrm{Re}\,ct(x/c).\sum_i (x-ib)\mathrm{Re}\,ct(x/a) \qquad (1)$$

Where c is the width of the grating transparent line, b is the pitch of the grating, and a is the length of the grating. Its Fourier transform consists of a series of spots aligned in the Fourier plane perpendicular to the grating lines, and it can be written as

$$F[f(x)] = F[\mathrm{Re}\,ct(x/c)]F\left[\sum_i (x-ib)\right].F[\mathrm{Re}\,ct(x/a)]. \qquad (2)$$

For an equal space-line grating, i.e. b=2c, its Fourier transform consists of Zero order and odd order spots only, the size of the spots is determined by the term F[Rect(x/a)] which is a sine function.

Fig. 3 shows the diffraction of a coherent optical Fourier transform system. At P_1 the transparency made by photographically copying the printed (projected) grating is placed. For real time application of DGN a spatial light modulator should be employed. A Cartesian co-ordinate system is chosen with its x, y plane in the Fourier Plane and its origin located on the optical axis. The FT-axis is the line which passes all the diffraction spots. In order to obtain high sensitivity and accuracy, the grating lines are placed in a direction approximately perpendicular to one of the coordinate axes, say x direction. The distance between two consecutive diffraction spots is given by

$$u = \lambda f / b, \qquad (3)$$

Where λ is the wavelength of the laser, f is the focal length of the FT lens, and b is constant of the grating.

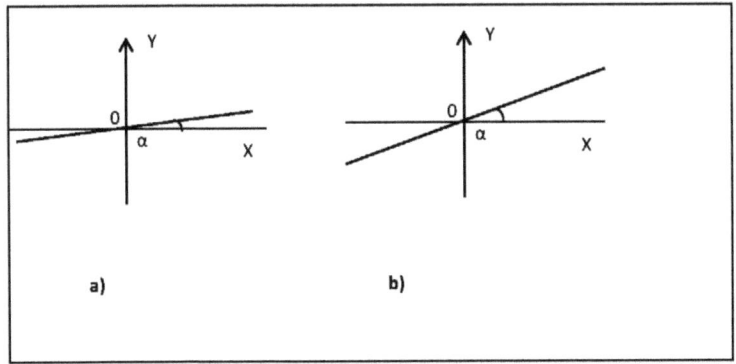

Fig. 4. Fourier transform of grating: a) before deformation, b) after deformation.

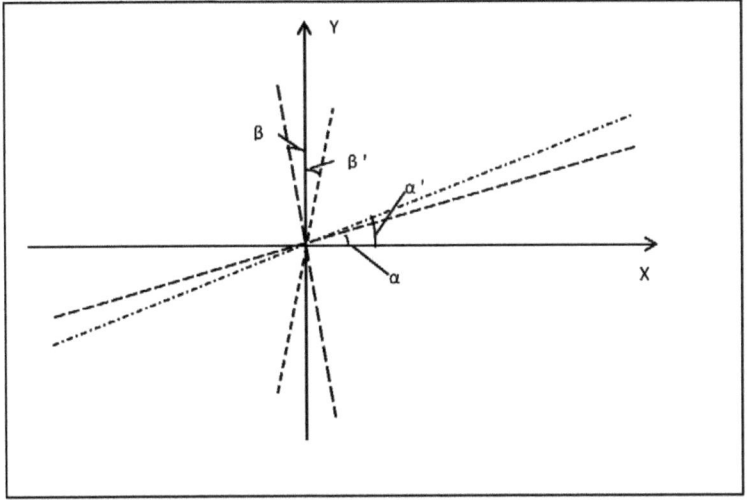

Fig. 5. Fourier transform of a grille.

Fig. 4. shows Fourier transform of an undeformed and deformed grating. From this figure the strain component in x-direction can be obtained by

$$E_x = (u\cos\alpha - u_x\cos\alpha')u_x\cos\alpha' \qquad (4a)$$

In the same way the strain components in y-direction can be given by

$$E_y = (u\cos\beta - u_y'\cos\beta')u\,y\cos\beta' \qquad (4b)$$

Naturally the grating lines should b parallel to the x-axis for E_y..

In order to find the corresponding angle deviation , the model surface must be provided with a grille instead of a grating. It is easy to see that the deviation can be obtained by

$$\gamma = (\alpha' - \alpha) + (\beta - \beta') = \alpha' - \beta'. \qquad (5)$$

So far has been shown that in plane (2-D) deformation can be determined completely by Fourier transform techniques. In the next section we will discuss 3-D deformation measurement.

3. 3-D deformation measurement

In case of 3-D deformations of a model the following changes will occur: the surface segment undergoes in-plane deformation and 3-D rotation.

As a matter of fact, we assumed in the last section that the transparency of the grating (or grille) was made by a camera whose axis is perpendicular to the surface segment. If this assumption does no hold true, what has been recorded on the transparency can be considered as the projection of the grating onto the image plane of the camera. In the same way we can consider the Fourier transform of the projected grating as the projection of the Fourier transform, as long as the optical imaging system is a telecentric system, and magnification remains constant.

For 3-D deformation measurement, the transparencies should be made from two different view angles, to be exact two transparencies are made before deformation (one for each view angle), another two transparencies will be made after deformation.

The Cartesian coordinate systems x,y,z and x',y',z', are chosen for undeformed and deformed state respectively. The x',y',z' system is rotated by the angle θ around the y-axis (fig. 5).

In order to determine the direction of FT-axis in 3-D space, the direction cosines are used. If the angle between the FT axis and the z-axis is ω_1, for a x,y,z system, they can be written as:

$$\sin \omega_1 \cos \alpha_1$$

$$\sin\omega_1 \sin\alpha_1 \qquad (6)$$

$$\cos \omega_1.$$

Similarly for the x',y',z' system, the direction cosines of the FT-axis are given by:

$$\sin \omega'_1 \cos \alpha'_1$$

$$\sin \omega'_1 \sin \alpha'_1 \qquad (7)$$

$$\cos \omega'_1.$$

The direction of the FT axis can be determined by knowing ω'_1 or α'_1. Applying the formulae for the transformation between coordinate systems, it is easy to show that the x',y',z' direction cosines can be obtained by the following matrix multiplication.

$$\begin{pmatrix} \cos\theta & 0 & -\sin\theta \\ 0 & 1 & 0 \\ \sin\theta & 0 & \cos\theta \end{pmatrix} \cdot \begin{pmatrix} \sin\omega_1 \cos\alpha_1 \\ \sin\omega_1 \sin\alpha_1 \\ \cos\omega_1 \end{pmatrix}$$

or

$$\begin{pmatrix} \sin\omega_1 \cos\alpha\cos\theta - \cos\omega_1 \sin\theta \\ \sin\omega_1 \sin\alpha \\ \sin\omega_1 \cos\alpha_1 \sin\theta + \cos\omega_1 \cos\theta \end{pmatrix} \qquad (8)$$

By combining eqs. (7) and (8) we get the following simultaneous equations

$$\sin\omega_1 \cos\alpha_1 \cos\Theta - \cos\omega_1 \sin\Theta = \sin \omega'_1 \cos \alpha'_1$$

13

$$\sin\omega_1 \sin\alpha_1 = \sin\omega'_1 \sin\alpha'_1 \qquad (9)$$

$$\sin\omega_1 \cos\alpha_1 \sin\Theta + \cos\omega_1 \cos\Theta = \cos\omega'_1,$$

where ω_1 and ω'_1 are unknown.

The distance between two consecutive diffraction spots is given by

$$u_1 = \left(\frac{u_{1xy}}{\sin\omega_1}\right) \qquad (10)$$

where u_{1xy} is the x,y plane projection of the distance u_1.

Since u_{1xy} can be measured, $\sin\omega_1$ can be obtained from eq. (9), u_1 is known. The above procedure can be repeated for the deformed grating to find u_2, the distance between two consecutive spots after deformation. Knowing u_1, u_2 and all direction angles enables one to determine the strain components in the x,y,z directions. Recall eqs. (4) and (5), the strain components will be given by

$$E_x = (u_2 \sin\omega_2 \cos\alpha_2 - u_1 \sin\omega_1 \cos\alpha_1)/u_2 \sin\omega_2 \cos\alpha_2$$

$$E_y = (u_2 \sin\omega_2 \sin\alpha_2 - u_1 \sin\omega_1 \sin\alpha_1)/u_2 \sin\omega_2 \sin\alpha_2 \qquad (11)$$

$$E_z = (u_2 \cos\omega_2 - u_1 \cos\omega_1)/u_2 \cos\omega_2.$$

A grating is used for the above analysis. In order to find the corresponding angle deviation necessary to print a grille instead of a grating. It should be pointed out that one needs to detect angle deviation for the 3-D situation. No matter how this angle is defined, it can be determined for the direction cosines obtained. When using a grille, we can find the direction cosines of two FT for each surface segment, so the orientation of the segment can be found. This may be helpful to interpolation technique for determining the deformation of the surface in positions where no grating (or grille) is printed.

4. Sensitivity of measurement

The sensitivity of the measurement is a great concern for every metrological technique. For the above analysis it can be shown that the sensitivity of measurement obtainable by the DGDM method is determined by the resolution of the optical systems (both imaging and Fourier transform system). An imaging system of high resolution allows to employ a grating (or grille) of high spatial frequency. The resolution of the Fourier transform system determines the size of the diffraction spots, i.e. the grating size can be used to estimate the smallest distance measurable on the Fourier plane. The theoretical limit of the measurable sensitivity can be estimated by

$$\delta = \lambda f/a \qquad (12)$$

Suppose 1) the size of the grating is a = 20 mm, 2) the focal length of the FT lens is 76 mm, 3)the wavelength of the laser source is chosen as $\lambda = 0.6328$ μm, then the sensitivity is given by

$$\delta = 0.6728 \times 760/20 \ \mu m = 24 \ \mu m,$$

and so the smallest strain which can be measured is

$$E = 24/19{,}237 = 0.125 \ \%.$$

The direction angles and the angle deviation can be achieved by measuring the distances on the Fourier plane. The measurement sensitivity of angles could be estimated from the corresponding equations. For simplicity, we estimated the smallest angle measurable for a simple situation. The granting is supposed to rotate by an angle ϕ around y axis, so the rotation can be considered to have same influence on the Fourier transform as the in-plane strain in x direction. Assume b' is the pitch projection onto the x, y plane, then we have

$$\phi = \cos^{-1}(b'/b)$$
$$= \cos^{-1}(1-0.125)$$
$$= 2.86^0.$$

5. Experimental results

The transparency obtained from a region of the imagery is placed on the set up and a grating has been used to perform FT transform (Fraunhoffer diffraction). During the experimental grating of 100 cycles/in was employed, the focal length of the lens is 760 mm, the wavelength λ of the He-Ne laser was chosen as 0.6328 μm. The displacements of diffraction spots were measured by using a scanning microscope mounted on an x, y translation stage. The experimental data are listed in table

Table 1. Experimental data

d_2	Δd_2	$\Delta d_2/d_2$ (%)	Displacement of diffraction spots	
			Experiment	Theoretical
650 mm	1 mm	0.15	10th 20th	10th 20th
650 mm	6.5 mm	1	182 342	162 324
650 mm	65 mm	10	1,726 3,017	1,619 3,238

Note : all the displacements are in μm.

6. Comments

In addition to our results the following comments could be made:

1. Better results can be expected by using a coherent optical system. Compared with the single-slit diffraction method, more light is available for easy observation.
2. Higher measurement sensitivity can be achieved either by observing diffraction spots of higher orders or by employing gratings of higher frequency and observation of lower order diffraction spots would offer both higher sensitivity and accuracy. The reason is that high linearity is obtainable when observation takes place near the optical axis of the FT system. During the experiment spots as high as the 40th order could be seen clearly with naked eyes , but we had hardly time to see even the 20th order spot with the scanning microscope which was being moved in a plan perpendicular to the optical axis. It was just because of the very large diffraction angle. It should be pointed out that a grating of 1,000 cycles/in could be printed on a model surface with suitable control of the processing , the diffraction properties of the grating could be controlled so that strong diffraction orders were produced in order number 10……25 [3].

Our proposed method is a robust one. It is not subject to environmental conditions such as vibration and diffusion, which are related to holographic interferometry and other interferometry methods. The required equipment for application of this method is reasonably cheap and has low weight and size. It is easy to obtain a very clear image of the object under investigation with the pattern for real time application of this method.

References

[1]   J.Apricio, J.L.Molpeceves, A.M.Frutos et al.: Improved algorithm for the
        analysis of Holographic Interferograms. Opt. Eng. **32**(1993) 963-969.

[2]   D.W.Robinson, G.T.Reid, Editors: Interferogram Analysis: Digital Pattern
        Measurement Techniques: Institute of Physics Publishing LTD 1993.

[3]   R.J.Collier, C. B. Bunkharott, L.H.Lin: Optical Holography. Academic Press,
        Newyork 1971.

[4]   C.M.Vest : Holographic Interferometry. John Wiley and Sons, NewYork 1979.

[5]   ASM Committee on Holographic Inspection (J.Romono et al. : Inspection by Optical
        Holography, ASM metals Handbook, Eight, Edition: Non-destructive
        and Quality control (1976) 198-222.

[6]   B.P. Hilderbrand: A General Analysis of Contour Holography. Ph.D. Diss.,
        Univ. Michigan 1976.

[7]   D.M. Meadows, W. O. Johnson, J.B. Allen: Application of Moire Contouring.
        Appl. Opt. **9** (1970) 942-949.

[8]   O.D. Soares: Hologram Repositioning by an Interferometric Technique. Appl.
        Opt. 18 (1979) 3838-3840.

[9]   Lord Rayleigh (Strutt, J.W.): Moire Fringes. Phil. Mag. **47** (1874) 81-193.

[10]   P.Beckmann, A. Spizzichino: The scattering of EM waves from rough
        surfaces. Pergamon press, Oxford 1963.

[11]   R.F.Wueker, L. D. Helflinger: Ruby Laser R1-R2 Contouring. Presented to
        the Conference and Exhibition on the Engineering uses of Coherent Optics,
        University of Strathclyde, Glasgow, Scotland, April 1975, 208-312.

[12]   J.Kokaj et al.: Optical Computation of Moments for Deformation
        Measurement. Optic **101** (1995) 49-52.

[13]   J.Kokaj, D.Casasent, Y.Q. Li: Coherent optical technique for transonic wind
        tunnel measurement. SPIE-Proc. **473** (1983) 23-28.

[14]   J.Kokaj, D. Casasent, J. Demorcopolow: A new statistical image processing for
        deformation measurement. Trends in Quantum Electronics, Section IV.
        Optical Processing of Information and Holography. Tom 34, 7-9 (1989) 961-965.

[15]   J.Kokaj et al. : An optical processing for Deformation. Measurement. Fizika
        22 (1990) 615-620.

[16]   D.Casasent, J.Kokaj: Morphological processing to reduce shading and
        illumination effects. SPIE-Proc. **1385** (1990) 152-163.

[17]   T.R. Pryor, W.P.T. North: Diffractographic strain gauge. Exper.Mech. **11** (1)
        (1971) 565-568.

[18]   P.M. Boone: Method for directly determining surface strain fields using
        diffraction gratings. Exper. Mech. **11** (1) (1971) 481-489.

[19]   C.A.Walker, J.Mckelvie: Practical Multiplied – Moire System. Exper. Mech.
        18 (1987) 316-320.

[20]   M.Tokeobr: Spatial – Carrier Fringe – Pattern Analysis and its Application, as
        Precision Interferometry and Profilometry: An overview. Industrial Metrology
        1 (1990) 79-99 .

[21]   M. Takedo, K. Mutoh: Fourier Transform Profilometry for the Automatic
        measurement of 3-D Object Shapes. Appl. Opt. **22** (1983) 3977-3962.

Section 2:

Optical computation of moments for biomedical deformation measurement

Optical computation of moments for deformation measurement. A new approach to deformation measurement is presented. It utilizes the moments of a pattern printed on the model surface under investigation .A coherent optical processor with a special mask is used to generate all the moments of a transparency made by photographically copying the printed pattern. A comparison of he moments computed before and after deformation enable to determine the sensitivity and discussed.

1. Introduction

Optical image processing, data processing have been attractive to researchers in various disciplines [1-7] .This has occurred because of the high speed , parallel processing , real time processing capabilities of such systems and because of their low size , weight , volume and cost .Hybrid processors that compute , invariant moments [8,9] for pattern recognition have received considerable attention [8, 13] .According to viewing area ($1.2 \ m^2$) of some metal parts of the car production, environment , and requirement for in plane and out – plane deformation measurement , make moment technique attractive .

Shown in fig. 1 is the block diagram of the moment estimator for deformation measurement.

We refer to our problem as distortion estimation, since the purpose of the processing is to determine in distortion in the input object. We consider the use of different observation spaces within which to view the object data. The premise is that the extraction of distortion parameters is easier if the object data is transformed into a new representation, other than the conventional (x, y) .Cartesian coordinate display .From these new observation spaces, we extract coefficients that describe the object. This set of values are the feature vectors, such are moments as a statistical concept.

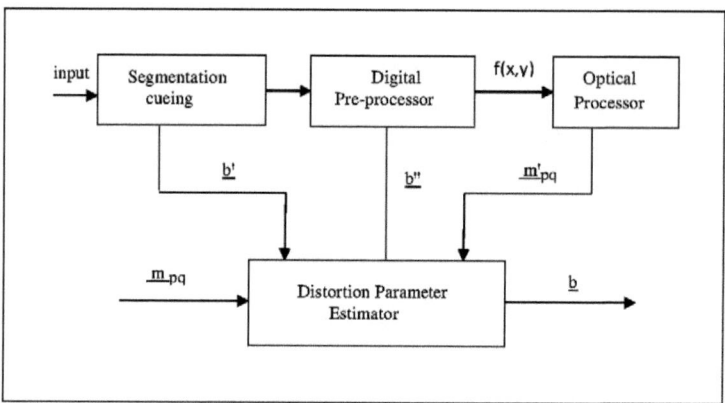

Fig. 1. Block diagram of the moment estimator for deformation measurement.

The following steps should be taken to compute the accurate distortion parameters, i.e , deformation:

1. Compute the moments m_{pq} of each pattern f (x, y) before deformation.
2. Compute the moments m_{pq} of each pattern f (x, y) after deformation
3. Comparing the moments m_{pq} with the moments m_{pq} to get the coars of initial estimates of the distortion parameters b_0.
4. Using iterative algorithm to optimize

$$b_j = b_{j-1} + \left(J^T \sum{}^{-1} J \right)^{-1} J^T \sum{}^{-1} \left[m - m_i(b_{j-1}) \right] \qquad (1)$$

where j is the iteration index and J is Jacobian $m_i(b)/b$ of the m(b) evaluated at b_{j-1}

5. Resulting interpolation techniques to get the deformation for the surface positions where no patterns are printed.

In this paper, we will concentrate on: how the moments change when the model undergoes deformation, the measurement sensitivity obtainable, some requirements for the coherent optical processor to compute moments compare to other coherent optical techniques [14-16].

2. Moments of a pattern function

For background purpose , we will first briefly review the definition of the moments , as a statistical concept pattern functions to find suitable patterns for our new techniques.

The definition of the moments of the function f(x, y) is

$$m_{pq} = \int f(x,y)x^p y^q dxdy \qquad\qquad (2)$$

The order of the moments is the sum p+q .Only lower order moments have physical significance. From mechanics , the combination of second order moments ($m_{02} + m_{20}$)$^{1/2}$ is the radius of the gyration of the function about the origin (this is related to the moment of inertia of the function)

The moments for a simple pattern function :

$$f(x,y) = \begin{cases} 1 & 0 \le x \le a \text{ and } 0 \le y \le b \\ 0 & otherwise \end{cases}$$

For a cosine grille function

$$f(x,y) = \begin{cases} 1 & (1 + \cos x)(1 + cosy)/4 \\ & 0 \le x \le a \text{ and } 0 \le y \le b \\ 0 & otherwise, \end{cases}$$

the moment expression can be written as follows:

$$M_{00} = a^2 b + \frac{b}{\omega_1}\sin \omega_1 a + \frac{a}{\omega_2}\sin \omega_2 b + \frac{1}{\omega_1}\sin \omega_1 a. \frac{1}{\omega_2}\sin \omega_2 b$$

$$M_{10} = \frac{1}{2}a^2 b + b\left[\frac{a}{\omega_1}\sin \omega_1 a + \frac{1}{\omega_1^2}(\cos \omega_1 a - 1)\right] + \frac{1}{2}a^2.\frac{1}{\omega_2}\sin \omega_2 b + \frac{1}{\omega_2}\sin \omega_2 b$$

$$.\left[\frac{a}{\omega_1}\sin \omega_1 a + \frac{1}{\omega_1^2}(\cos \omega_1 a - 1)\right] \qquad\qquad (3)$$

$$M_{01} = \frac{1}{2}ab^2 + \frac{1}{2}b^2 \frac{1}{\omega_1}\sin \omega_1 \, a + a \cdot \left[\frac{b}{\omega_2}\sin \omega_2 \, b + \frac{1}{\omega_2^2}(\cos \omega_2 b - 1)\right]$$

$$M_{11} = \frac{1}{4}a^2b^2 + \frac{1}{2}b^2\left[\frac{a}{\omega_1}\sin \omega_1 a + \frac{1}{\omega_1^2}(\cos \omega_1 a - 1)\right] +$$

$$\frac{1}{2}a^2\left[\frac{b}{\omega_2}\sin \omega_2 \, b + \frac{1}{\omega_2^2}(\cos \omega_2 b - 1)\right] + \left[\frac{a}{\omega_1}\sin \omega_1 a + \frac{1}{\omega_1^2}(\cos \omega_1 a - 1)\right]$$

$$\cdot\left[\frac{b}{\omega_2}\sin \omega_2 \, b + \frac{1}{\omega_2^2}(\cos \omega_2 b - 1)\right] \quad\quad (4)$$

Under some assumption and calculations the final simplified moment expressions will be

$$m_{pq} = a^{(p+1)}b^{(q+1)}/ \;^{(p+1)(q+1)} \quad\quad (5)$$

where $a=b=1$ are the assumed dimensions for a simple pattern.

3. Change in moments and deformation

When the model undergoes some deformation, the printed pattern will be distorted correspondingly, therefore the moments of the pattern will change. Changes in moments enables one to determine the model deformation . Both in – plane deformation and out-of-plane deformation will be discussed on the assumption that the pattern undergoes uniform deformation

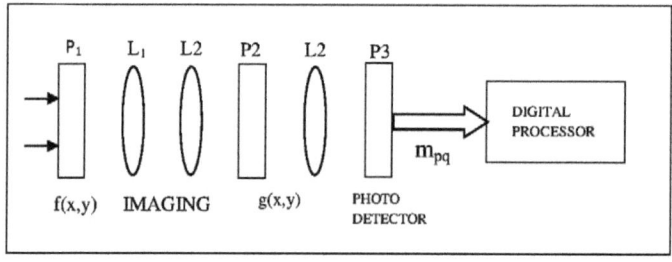

Fig. 2. Coherent optical processor.

3.1 In Plane deformation

We will discuss in – plane scaling, the most common and practical deformation .Assume

1) the dimension of the pattern a=b=1

2) the scale fraction in x-direction is and the scale factor in y-direction is then moments of the scaled pattern can be written as

$$m_{pq} = a^{(p+1)} \beta^{(q+1)} / (p+1)(q+1) \qquad (6)$$

the relative changes in moments can be written as

$$c_{pq} = \frac{m_{pq}}{m_{pq=a}^{p+1}} \beta^{(q+1)}. \qquad (7)$$

Either factor a or β can be found simply using

$$a = (c_{p0}/c_{00})^{1/p} \qquad (8)$$

$$\beta = (c_{oq}/c_{00})^{1/q} . \qquad (9)$$

3.2 Out –of –plane rotation

Suppose the pattern rotates by an angle about y-axis . As long as a telecentric imaging system is used to make the transparency of the rotated pattern , the transparency can be considered as projection of the original pattern , and rotation angle can be given by .

$$c_{p0}/c_{00} = (\cos\phi)^p \qquad (10)$$

ie.,

$$\phi = \cos^{-1}(c_{p0} / c_{00})^{1/p} . \qquad (11)$$

The rotation angle about x-axis could be obtained in a similar way.

4. Requirements for the optical processor

Fig.2 shows the coherent optical processor for computing the moments of an input pattern. The pattern is the transparency made by photographically copying the printed pattern , which is placed in p_1 and imaged into p_2 ,where a special mask is

23

placed as well. The light leaving p_2 is thus a product $f(x,y)\,g(x,y)$. The Fourier transform , made by lens L_3 , is measured in plane P_3 by a photodetector .

There are three important requirements for the optical processor :

1.Photodetector dynamic range.

2. mask spatial resolution.

3. mask gray level quantization.

4.1 Required detector dynamic range

It can be easily shown that moment depend greatly on either the selection of the co-ordinate origin of the direction of the co-ordinate axis. If we assume that $a = b = 1$, the moments for the square pattern will decrease as the order of the moments goes higher. The ratio of m_{00}/m_{pq} determines the required dynamic range of the detector. We know from (3) that $m_{00} = 1$ and $m_{10,0} = 1/11$, obviously, the dynamic range of commercial available detector is much more larger than the above requirement. The more decisive factor is the measurement sensitivity. The smallest detectable moments are limited by the detector sensitivity.

4.2 Required mask spatial resolution

In practice, the mask to be used for computing moments optically would be produced by using a computer controlled film recorder. Instead of a continuous mask function, a discrete mask function is recorded on a film plate, i.e., the mask function is digitized both spatially and in amplitude (gray level). Digitization of the spatial coordinate will be referred to as mask sampling, while digitization in amplitude will be called gray level quantization.

For the simple square pattern, we know from equation (3) that the moment expression can be written as $m_{pq} = a^{(p+1)}b^{(q+1)}/\,{}^{(p+1)(q+1)}$. Mask sampling will cause some error in either x or y coordinate measurement, namely in the measurement of a and b. Suppose the sampling interval is Δ_s , then the error in m_{pq} caused by the sampling is given by

$$m_s = m_{pq}/m_{pq}\,\overline{\Delta}_s(p + q - 2).$$

Intuitively, the sampling interval Δ_s should be much smaller than the possible change in pattern dimensions.

4.3 Gray level quantization

When moments are computed optically by using a coherent optical processor with a special mask, the definite continuous integrations are approximated by numerical integrations. Fig.3 shows the numerical integration of a 1- D function $y = x^p$.

The moments of a 1- D shifted rectangular function are given by

$$m_p = \int_0^1 x^p dx = x^{(p+1)}/(P+1)|^1_0.$$

Its numerical version is

$$m_p = \Delta x(x_0^p + x_1^p + \dots \dots \dots + x_n^p - 1).$$

Because of gray level quantization, there is an uncertainty in measurement of x^p; the x^p are in error by at most g (gray level quantization)

$$Mp = \Delta x.n.\Delta g$$

i.e.,

$$m_p/m_p = g(p+1).$$

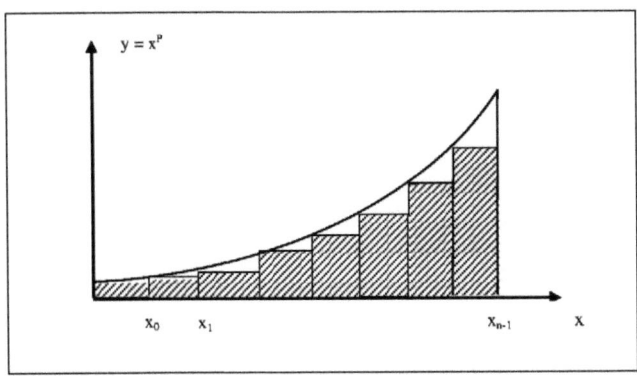

Fig. 3. Numerical integration of a 1-D function $y = x^p$.

25

For a 2-D function, the result can be extended to find the error in moments. It can be shown that relative change in moments 2-D function is given by $m_g m_{pq}/m^{pq} = g(p+1)(q+1)$. It should be pointed out that this error is the difference between the theoretical moments and the moments computed optically using the digitized mask; it remains a constant as long as no deformation occurs. Recalling the way we derive the above expression, the change in this constant error caused by deformation can be given by,

$$m_g = \Delta g(p+1)(q+1) - c_d$$

where c_d is a multiplication factor which is determined by deformation which is determined by deformation. When the pattern undergoes scaling factor in x and y direction, respectively $c_d = 1 - \alpha \beta$.

5. Summary and conclusion

A statistical image processing has been proposed for deformation measurement. A pattern is printed on the model surface under investigation, a transparency of the pattern is placed at the input plane of a coherent optical processor with a special mask to compute all the required moments of the pattern. Changes in moments enable one to determine the deformation. The requirements for the optical processor were discussed. The results show that the mask sampling sets a more critical limit to the implementation of the system than the gray level quantization requirement does. By using equations (3), (4), (5), (6) numerically it can be shown that the scale factor is $\alpha = (0.99)^{1/10} - 0.999$ and the out-of-plane rotation angle $\phi = \cos^{-1}(0.999) = 2.56^0$. These are the results for scale change (0.1 %) and rotation angle detected by our new method.

References

[1] B.K.P. Horn: Image Intensity Understanding Artificial Intell. **8** No. 2 (1977) 201-231.

[2] Y.Nakagawa: Automatic Visual inspection of Solder joints on printed circuit boards. Proc. SPIE Robot Vision **336** (1982) 121-127.

[3] D.Rumelhart, D.Zipser: Feature discovery by competitive learning. Cognitive Science **9** (11) (1985) 75-112.

[4] D.Casasent. Optical Morphological Processors. Proc. SPIE **1350** (1990) 380 - 394.

[5] J Kokaj, D. Casasent, R. Schaefer : Morphlogical Processing to Reduce Shading and Illumination Effects. Proc. SPIE **1385** (1990):152-164.

[6] D. Casasent, R. Schaefer, J.Kokaj: Morphological processing to determine statistical size and direction distributions. Intelligent Robots and Computer Vision-Image Processing **9** (1990) 1381-1310.

[7] J.Kokaj, Y.Makdisi, K.Bhatia : Biomedical image analyis using Morphology processing . Opt.Eng. (accepted for publication).

[8] M.K.Hu: Visual Pattern recognation by moment invariants. IEEE Trans. Inf.Theory **IT-8** (1962) 179- 182.

[9] G.B. Gurevich : Foundations of the theory of Algebraic Invariants. Publisher P.Noardhoff Ltd, Groningen, The Netherlands 1964.

[10] D.Casasent, D.Psaltis: Optical pattern recognation using normalied Invariantsmoments Proc. SPIE **201** (1979) 107 – 194.

[11] D.Casasent, A. Furman: Source of Corelations Degradation. Appl. Opt. **16** (1977) 1652 – 1661.

[12] D.Casasent,D.Psaltis: Hybrid processor to compute Invariant moments for Pattern recognation. Opt. Lett. 5 (1980) 359 – 397.

[13] S.Dudani et al.:Aircraft identification by moment Invariants. IEEE Trans. Comput. **26** (1977) 39-46.

[14] J.Kokaj, A.Dida: Proc.SPIE 473 (1984) 87-90.

[15] J.Kokaj, Li Yong-Qing : Experimentelle Technik der physik 37 (1989) 23-28.

[16] J.Kokaj, Y.Makdisi, K.Bhatia: A new Optical approach for deformation \ measurement. OPTIKA pura y Applicada, (1994) 27 (3) 175.

Section 3:

Higher-moment and diffraction-order computation for deformation measurement

Higher moment and diffraction orders are computed and used for deformation measurement. For a given two-dimensional function lower-order moments are calculated. It is shown that simplified results can be obtained if an appropriate pattern is used for the calculation. The obtained results are employed for calculation of higher-order moments. Optical implementation and their use for deformation measurement is shown. High diffraction orders are introduced by using a projected grill-pattern on the model surface. This is performed by selecting higher frequency spots in the Fourier-Transform plane. Selection of the high-frequency diffraction orders and calculation of high-order moments are two image processing approaches useful for deformation measurement. Accuracy of the measurement for in –plane deformation and out-of-plane rotation is calculated. It is shown that accuracy of measurement is increased when higher orders are used.

1. Introduction

The recent advancements in spatial light modulator technology [1-4] and optical linear algebra processors have introduced new approaches in optical computing and optical image processing. The parallel processing, high speed, compact system fabrication, low-power dissipation and size of optical processors have achieved great strides in recent years. The architecture's algorithms and system fabrication of hybrid processors have made them robust and attractive techniques for industrial application [5, 6], pattern recognition [7,8], military application [9] and for deformation measurement [10-12]. Moments as a new feature space [13] has been used for pattern recognition [14-16]. In this paper a hybrid architecture for computation of higher-order moments, higher diffraction orders and their use for deformation measurement are proposed. In Sections 2-6 of this work an advanced approach is proposed where high order moments are introduced. In sections 6 and 7 the diffraction on a grill as an approach for deformation measurements is shown. Compared to other approaches [17-23] the simplicity and accuracy of this one is shown. We will show that higher diffraction orders are also useful for accurate and sensitive measurements.

2. Higher-order moments

In order to calculate higher-order moments, and to demonstrate their use in deformation measurement later on, brief analysis of this concept is presented. The moments m_{pq} of a function $f(x, y)$ are defined as in equation (1) of Section 2.

The order of the moments is the sum $p+q$. The lower-order moments have some physical significance. The zero-order moment m_{00} corresponds to energy of the function, whereas the first- order moments m_{01} and m_{10} give the centroid of the function in the directions x and y, respectively. In mechanics, the combination of the second-order moments $(m_{02} + m_{20})\exp[1/2]$ is the radius of gyration of the function about the origin and this is related to the moment of inertia.

The function $f(x, y)$ in Eq.(1) represents the pattern which can be chosen according to the task of the moment application. Here we choose the function $f(x, y) = 1 + \cos\omega_1 x + \cos\omega_2 y + \cos\omega_1 x \cos\omega_2 y$. Assuming that a and b are the extends of the pattern in x and y direction as a continuation of lower order moment shown in Section 2, higher order moments are computed as

$$M_{21} = \frac{1}{6}a^3 b^2 + \frac{1}{2}b^2 \left[\frac{a^3}{\omega_1}\sin\omega_1 a + \frac{2a}{\omega_1^2}\cos\omega_1 a - \frac{2}{\omega_1^3}\sin\omega_1 a\right] + \frac{1}{3}a^3 \left[\frac{b}{\omega_2}\sin\omega_2 b + \right.$$

$$\frac{1}{\omega_2^2}(\cos\omega_2 b - 1)\right] + \left[\frac{a^2}{\omega_1}\sin\omega_1 a + \frac{2a}{\omega_1^2}\cos\omega_1 a - \frac{2a}{\omega_1}\sin\omega_1 a\right]$$

$$\left[\frac{b}{\omega^2}\sin\omega_2 b + \frac{1}{\omega_2^2}(\cos\omega_2 b - 1)\right] \tag{1}$$

$$M_{12} = \frac{1}{6}a^2 b^3 + \frac{1}{3}b^3 \left[\frac{a}{\omega_1}\sin\omega_1 a + \frac{1}{\omega_1^2}(\cos\omega_1 a - 1)\right] +$$

$$\frac{1}{2}a^2 \left[\frac{b^2}{\omega_2}\sin\omega_2 b + \frac{2b}{\omega_2^2}\cos\omega_2 b - \frac{2}{\omega_2^3}\sin\omega_2 b\right] + \left[\frac{b^2}{\omega_2}\sin\omega_2 b + \frac{2b}{\omega_2^2}\cos\omega_2 b - \right.$$

$$\frac{2}{\omega_2^3}\sin\omega_2 b\right]\left[\frac{a}{\omega_1}\sin\omega_1 a + \frac{1}{\omega_1^2}(\cos\omega_1 a - 1)\right] \tag{2}$$

For simplicity, the remaining calculated orders are not included.

3. Simplified version of the moment expression

If the spatial frequency of the pattern is few lines/mm, it is easy to show that all terms containing sine function will vanish, and the terms containing cosine function will be one. Therefore, the simplified version of the moment expressions will as follows:

$$M_{00} = ab, \qquad\qquad M_{01} = \frac{1}{2}ab^2,$$

$$M_{10} = \frac{1}{2}a^2b, \qquad\qquad M_{20} = \frac{1}{3}a^3b + \frac{2ab}{\omega_1^2} \qquad\qquad (3)$$

For simplification other calculated moments are not included in the following are

$$M_{30} = \frac{1}{4}a^4b + \frac{3a^2b}{\omega_2^2}, \qquad M_{21} = \frac{1}{6}a^3b^2 + \frac{ab^2}{\omega_1^2} \qquad\qquad (4)$$

$$M_{03} = \frac{1}{4}ab^4 + \frac{3ab^2}{\omega_2^2}, \qquad M_{12} = \frac{1}{6}a^2b^3 + \frac{a^2b}{\omega_2^2}$$

By comparing (2) – (5) with (6) – (7) it is easy to see that inside structure of the pattern does not have much influence on the moments. So the lowest frequency pattern can suffice. By introducing the pattern.

$$f(x,y) = \begin{cases} 1 \; 0 \le x \le a, \\ 0 \le y \le b, \\ 0 \; otherwise, \end{cases} \qquad\qquad (5)$$

the following moments are calculated:

$$m_{00} = ab \,, \; m_{11} = \frac{1}{4}a^2b^2$$

$$m_{10} = \frac{1}{2}a^2b, \; m_{30} = \frac{1}{4}a^4b$$

$$m_{01} = \frac{1}{2}ab^2, \; m_{03} = \frac{1}{4}ab^4 \qquad\qquad (6)$$

$$m_{20} = \frac{1}{3}a^3b, \; m_{21} = \frac{1}{6}a^3b^2$$

$$m_{02} = \frac{1}{3}ab^3, \; m_{12} = \frac{1}{6}a^2b^3$$

By using the last simplified expressions, one can write any order of the moments.

4. Computation of the higher order moments using digital and optical processor

In Fig.1. of Section 2,a block diagram is shown for the estimation of moment, for the deformation measurement. To obtain the accurate distortion parameter – deformation, the following steps are proposed:

1. Compute the higher order moments m_{pq} of each pattern $f(x,y)$ before deformation.

2. Compute the higher moments m'_{pq} of each pattern $f(x, y)$ after deformation.

3. Comparing the higher moments m_{pq} with the moments m'_{pq} to get the initial course or initial estimates of the pattern distortion parameter b_0, as shown in eq. (1) of Section 2.

As can be seen in Section 2, an optical processor is required to compute the moments automatically and in parallel. The optical processor is shown in Fig. 2. On an object under investigation, a pattern is projected and after taking a picture a transparency is made. The obtained pattern is considered as the input function $f(x,y)$. Using the same procedure a new transparency is made after the object under investigation is deformed and it is represented with $f_d(x,y)$.

The function $f(x,y)$ from the plane P_1 is projected on to P_2 by passing a parallel light beam through the transparency P_1 and the imaging system. Before this projection a transparency P_1 and the imaging system. Before this projection a transparency representing the function $g(x,y) = x^p y^p$ is placed in the plane P_2.This transparency is actually the mask for optical computation of the desired orders of the moments by choosing values of p and q.By passing the parallel incident beam through the planes P_1 and P_2, the information from both transparencies are collected and are projected on to the plane P_3 by using a Lens L_2. In this way the moments m_{pq} (representing underformed state) are optically computed and projected on to the plane P_3. The same procedure is repeated for the deformed case.

By comparing the moments obtained before and after the deformation, the measurement of the deformation can be achieved.

Digital and optical algorithms used for calculation of moments, of Section 2. are used here for calculation of higher order moments as well.

5. High-order moments and deformation measurement

There are several experimental problems one should consider for optical computation of the higher moments and their use in deformation measurement. Here we consider two relevant problems: Measurement sensitivity and dynamic range of the photo detector used for moment detection shown on the plane P_3, in Fig. 2.

Measurement sensitivity can be calculated by introducing a simple pattern function which is prepared photographically and its transparency is placed in the plane P1. For function shown in Section 2, which represents a simple pattern, the corresponding moments are

$$m_{pq} = a^{(p+q)} b^{(q+1)} / (p+1)(q+1) .$$

According to Section 4, the printed pattern is destroyed when the object under investigation is deformed. Therefore, the moments of the pattern will change in case of in-plane and out-of-plane deformation.

Assuming that pattern dimension is a = b = 1, we consider the in-plane deformation or in scale factors in x and y directions. The moments for the deformed case are shown in eq . (6) of Section 2.

The factors α or β can be found by the eq. (8, 9) of Section 2.

Now supposing a scale change in the x direction we have

$$m'_{00} = \alpha^2,$$

$$m'_{30} = 1/4\alpha^5 ,$$

where the relative changes in m_{00} and m_{30} are

$$m'_{00}/m_{00} = \alpha^2 \qquad \text{and} \qquad m'_{30}/m_{30} = \alpha^5.$$

Assuming that the SNR of the detected signal is 40 dB, ie. 1/100 change in moment occurs, we obtain from m'_{00}/m_{00},

32

$\alpha = (0.99)^{1/2} = 0.9949874.$

From m'_{30}/m_{30} one obtains

$\alpha = (0.99)^{1/5} = 0.9979919$

and from m'_{90}/m_{90} the value is

$\alpha = (0.99)^{1/10} = 0.9989954.$

In the above cases, the scale changes of 0.005, 0.002, and 0.001 respectively, were obtained therefore, higher-order moments offer higher measurement sensitivity.

The next problem to be considered here is the dynamic range of the detector. The ratio m_{00}/m_{pq} determines the required dynamic range of the detector. We know from (11) that $m_{00} = 1$ and $m_{10} = 1/11$. Obviously, the dynamic range of commercially available detectors is much larger than the requirement.

6. Higher diffraction orders and deformation measurement

Higher diffraction orders are obtained from the projected grating on the model surface. A transparency is made by photographically copying the model and the projected (printed) pattern. We made a transparency from a projected grill on a surface which is placed in front of a collimated laser beam. The Fraunhoffer diffraction or Fourier transform (FT) is obtained by introducing a bi-convex lens behind the grating. The 2-D positions and inter-distances of the diffraction spots on the screen, obtained on the FT-plane provides information on deformation of the object under investigation. Accordingly, the changes on the object due to deformation, change the constant of the grill. Therefore the inter distance of the diffraction spots on the screen is changed. The experimental set up is very simple and is shown in Section 2.

The obtained transparency, as described above, is placed in front of a collimated laser beam, a lens of focal length f = 760 mm is placed behind the transparency(the grill). In the FT plane diffraction spots in x and y direction are obtained. A x-y scanning microscope is placed on the FT plane to measure the inter distances of the diffraction spots. The distance between two consecutive diffraction spots can be calculated by solving Fraunhoffer's diffraction problem using Kirchoff's [24] integral or by calculating Fourier transform of a two dimensional equally space line grating f(x,y) and is given by

$$u = \frac{\lambda b}{f},$$

where λ is the wavelength of the laser, f is the focal length of the FT lens, and b is the pitch of the grating. Since the size of the pitch is the same in the horizontal and vertical directions of the grill, we write a one dimensional function – a space line grating is demonstrated in Section 2. In case of the deformation of the actual body, encoded in a grill and placed on the input plane of the experiment, directions of the FT spots are rotated by angels $\alpha - \alpha'$ and $\beta - \beta'$ for the two directions, respectively. This can be seen in in Section 2. Therefore, by using the corresponding angle of deviation, for a grill pattern projected on the model surface it is easy to see that the deviation angle can be obtained by

$$\gamma = (\alpha' - \alpha) + (\beta' - \beta) = \alpha' - \beta',$$

where α and β are angles of the direction of the collinear diffraction spots before deformation-rotation, and α' and β' are the corresponding angles after deformation.

Two Cartesian coordinate systems are chosen, one for each view angle. One of the coordinate systems is named X, Y, Z – system, and the other one X', Y', Z' – system. They are shown in Fig. 1.

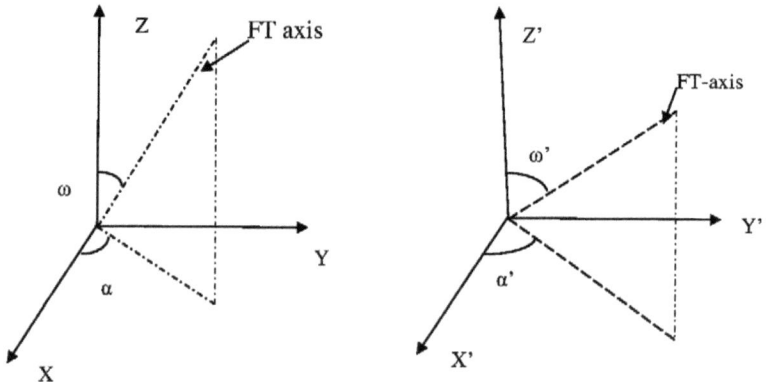

Fig.1. Projection of the FT before and after deformation

This technique is capable of determing the deformation of the parts-surfaces where a grill is printed. This can be done by using interpolation technique.

The measurement sensitivity is determined by resolution of the optical system and the size of the employed grating, denoted by α. The theoretical limit of the measurement sensitivity could be estimated inthe way shown in Section 2.

7. Experimental results

The aim of this experiment is to demonstrate one-dimensional deformation measurement using FT of the grating. The experimental set up is like the one shown in Section 2. with a small change: the grating is displaced from the position shown and is placed on the right side of the lens.The initial position of the grating from the FT plane is denoted by d2, The displacement of diffraction spots were measured by using a scanning microscope mounded on an x,y translation stage.

The focal length of the employed grating was 100 cycle/in , the focal length of the lens 760 mm , and the wavelength of laser was 0.6328 mm.

The experimental results obtained are shown in Table 1 of Section 2.

8. Conclusion

In this work, moments and diffraction orders are selected and used for deformation measurement. In both cases higher orders are selected .Higher –order moments using optical processor are computed .Inspite of the fact that high –order moments have no physical meaning, they can be applied in deformation measurement when a high accuracy is required. The method is simplified due to the fact that high frequency of the pattern projected on the object under investigation does not effect the momentcalculation .Therefore , a simple pattern like a square can be used.The dynamic range and other requirements of the mask are also important factors which will be considered in the future work. The use of moments requires a mask preparation of the function $x^p y^q$, which can be written in a transparency by using a computer –controlled laserbeam.For the FT-application a prepared grill is projected in the object under investigation.The interdistance of the diffracted spots obtained after the grating (transparency containing information of object and its deformations) which is employed on the diffraction setup, gives us information for actual deformation.High diffraction orders give us a better accuracy .We observed up to 40th diffraction order. However ,we used only 20th order ,because very high orders are not reliable for obtaining good result as they are nor clear and are not easy to be measured . Both methods are simple and robust for use in industrial application.They are not subject to vibration, or some other environmental restrictions.

References

[1] Fukushina S, Kurokawa T, Ohno M. Ferroelectric liquid-crystal spatial light modulator achieving bipolar image operation and cascadability. Appl Opt 1992;31:6859-68.

[2] Row MG, Scheher K. High-speed and high-contrast operation of ferroelectric liquid crystal optically addressed spatial light modulators. Opt Engng 1993;32:1662-7.

[3] Moddell G, Johnson K, Li W, Pagano-Strauffer L, Hadschy M. High-speed binary optically addressed spatial light modulator. Appl Phys Lett 1989:55:537-9.

[4] Weeks AR, Myler HR, Emery JD. Nonlinear image transformations implemented with spatial light modulators. Opt Engng 1994:33(3): 850-5.

[5] Botha E, Richards J. Casasent D. An optical laboratory morphological inspection processor. Appl Opt 1989: 28:5342-50.

[6] Casasent D. Coherent optical pattern recognation: a review special issue on Optical Computing. Opt Engng 1985;24:26-32.

[7] Casasent D. Iyer A. Ravichandran G. In plane distortion reconnaissance filters. SPIE 1990; 1342; 152-63.

[8] Casasent D, Shafer R, Kokaj J.Morphlogical processing to reduce shading and illumination effects. Proc. SPIE 1990;1385:152-64.

[9] Kokaj J, Li Young Quing. Deformation measurement using fourier transform technique. Exp Tech Phys, 1988;37:23-9.

[10] Kokaj J , Casasent D, Li Young Quing , Demakopolou J.A new approach of optical data processing for deformation measurement. FIZIK A A 1990;22(4):615-9.

[11] Kokaj J ,Dida A.Deformation measurement using correlation shape. SPIE, OPTIKA1984;84(437):235-9.

[12] Kokaj J , Casasent D, Li Young Quing .Some coherent techniques for deformation measurement. SPIE Optika 1984:84(437):244-9.

[13] Fukunaga K. Introduction to statistical pattern recognation.New York: Academic Press, 1972.

[14] Casasent D, Cheatham RL ,Proc ASME 1984;83:13-22.

[15] Casasent D, Cheatham RL Proc SPIE 1984;504:4-16.

[16] Kokaj J, Makdisi Y, Bhatia K .Optical moments for deformation measurement.Optik 1995;101(9):49-52.

[17] Zhan Gtao, et al.High speed fringe analysis by using stair-shaped virtual grating demodulation.Opt lasers Engng 1997;28:411-22.

[18] Tang J, Zhang H.Study on in–plane displacement measurement under impact loading using digital speckle pattern interferometry.Opt Eng 1996;35:1080-3.

[19] Koufman GH. Transient in–plane deformation analysis by means of pulsed TV holography. Optik 1998;108(1):43-7

[20] Kokaj Y, Galanger NC. Relative passes of electromagnetic waves diffracted by a perfectly conducting rectangular grooved grating. J. Opt. Soc. Am. 1988; 5:65-73.

[21] Forno C.Deformation measurement using high resolution moire photography. Opt lasers Engng 1988;8:179-212.

[22] Luo PF, Chao YJ, Sutton MA, Peters WH. Accurate measurement of three dimensional deformations in deformable and rigid bodies using computer vision. Exp Mech 1993;33:123-32.

[23] Suttan MA, Tuner JL, Bruck HA, Chae TA. Full-field representation of discreetly sampled surface deformation for displacement and strain analysis. Exp Mech 1991;31:168-77.

[24] Born M, Wolf, Principles of optics, 6th ed. Toronto, New York: Pergamon Press, 1993.

Section 4:

Deformation measurement of a human back using Moiré Technique

Simulated deformation of a human back is measured using moiré technique (MT). The reproducible positions of human body are fixed by using two scales for measurement of the weight and by measuring the distance of the tip of left hand from the left knee. Shadow and projection moiré are performed on the body for each position. The actual measurement of the deformation is performed by using projection moiré approach. Fringe interpretation of the obtained moiré figures is performed by using " Quick Fringe" software (QFS). The programs are modified according to the task of the experiment. Contour, 1-Dimensional and 2-Dimensional representations of the surface of the deformations is obtained for four selected positions of the human back. MT and fringe interpretation we propose, is associated with an actual clinical technique applied for diagnosis of ankylosing spondylitis , back strain related to primary osteo-arthritis and knyphosis. The possibility of using the gratings with changeable transparency is indicated.

INTRODUCTION

The main goal of this paper is to study problems of a human back using a simple and useful optical technique. The majority of causes of the back problems are due to prolapsed lumbar intervertebral disc and osteoporosis-arthritis of the spine. Scoliosis is deformity or curvature of the spine. Structural scoliosis is caused by deformity of vertebrae. However, if the vertebrae are normal, the deformity can be caused by other reasons. It may be compensatory, resulting from tilting the pelvis from real or apparent shortening of one leg. It may be static due to unilateral protective muscle spasm, especially accompanying a prolapsed intervertebral disc. In structural scoliosis there is alternation in shape and mobility of spine constituting a deformity that cannot be corrected by alternation of posture. A careful examination is require to find a cause and suggest prognosis [1-2].Structural scoliosis may be congenital, deformity being due (for example) to a semi vertebra (only half of a single vertebra is fully formed, fused vertebra of absent or fused ribs). In paralytic scoliosis the deformity is secondary to loss of supportive action of trunk and spinal muscles, nearly always as a sequel to anterior polimeliomylitis. A cause for scoliosis could be due to disorders of the supportive musculature of the spine as well. Common cases of scoliosis leading to spinal deformity are accompanied with shortening of the trunk, and there is often impairment to respiratory and cardiac function [3-4]. This could lead even to invalidism and shortening of life expectancy.

Deformation of the back by spondylitis is due to progressive ossification of the joints of the spine. The joints between D_{12} and L_{11} are often affected, but the rest of the thoracic and lumbar spine is rapidly involved with striking loss of mobility. Stiffness of the back and pain are persisting symptoms. Tuberculosis of the spine is another cause of the back deformation and may produce angular deformation called kyphotic or scoliotic deformities. Beside the clinical observation, x-ray examination is the next procedure for investigation of a patients back. The aim of this project is to introduce an optical nondestructive and noncontact technique to detect and identify the deformations of the back. The source of the apparent deformation of the back can be due to biomechanical deformities and muscle spasm.

In Fig.1. a human back and organs at the region of the back are shown. A pain of an internal organ can induce deformation of the back. While x-ray examination can identify the deformations of non transparent organs such as spinal cord, optical techniques such as moiré can detect and quantify the deformation of the skin, deformations caused by internal organs or other biomechanical problems of the body.

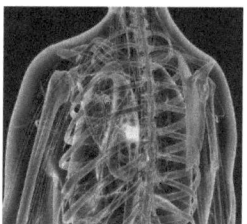

Fig.1. Human back and organs at the region of a human back [5] (Sebastian *et al.*, 2011).

The advantage of the optical techniques, in this case moiré technique, is detection of in-plane and out-of-plane deformations. This is a robust, economical and simple technique that could be applied during the initial examinations of a patient. This technique is based on optical fringe formation and interpretation. However, it does not require anti-vibration free platform as in the case of holographic interferometry. Moiré approach proposed is based on harmless electromagnetic waves of visible region [6,7, 8].

Deformation measured by this approach gives us valuable quantitative information based on the fringe interpretation Taksaki [9] in1970 has proposed, for the first time, a clinical application of Moire' phenomenon using shadow contour moiré technique. Since then there are many papers published on this topic.

Later onTaksaki [10], has improved his shadow-based contour Moire' by increasing the fringe contrast. The most comprehensive work related to medical applications of Moire' is one chapter of the book that can be found on Web in pdf form at the following address:*www.intechopen.com/download/pdf/36611*. The title is:"Moiré Topography: From Takasaki Till Present Day - InTech", written by Flavia 2012) [11](.More than 80 authors and citations can be found at this chapter of the book. Here authors have elaborated thoroughly the Shadow-Moire' Technique. The projection moiré is only mentioned with a picture obtained experimentally. The Moire technique has been successfully used for medical applications in the past [12, 13, 14, 15, 16, 17].In this paper we have shown comprehensively, deformation measured of the human back: we have obtained the images and performed fringe interpretation using "Quick Fringe". As to our best knowledge, this software is used for the first time for fringe interpretation, when MT is applied for biodiagnosis of ankylosing spondylitis, back strain related to primary osteo-arthritis, and knyphosismedical deformation measurement. The MT measurement have limitations on detection of the mentioned deformations and health problems.

The advantage of our proposed method, capable to obtain simultaneously1-D, 2-D and 3-D representation with numerical data, makes our proposed technique useful. The software has possibility of applying so called horizontal and vertical centers, or they can be applied simultaneously. We have shown that the use of only horizontal centers leads us to optimal output. As is indicated by many authors, the most important element used in MT application is the optical grating. In case of shadow moiré, only one grating is used. However, in case of projection moiré two gratings are used: object and reference grating. We have made and applied a new grating with linear changeable transparency.

While reading more than 70 references, we have notice that only static position and symmetry of contours have been analyzed by MT.

41

Here, we have studied several positions of the human body, simulating the dynamics of the motion. In collaboration with medical experts and by consulting books such as Clinical orthopedic examination [15] and Management of the Back Pain we have studied critical positions of the back of the patients. Beside clinical touch, doctors have used a flexible meter to obtain quantitative information to determine the deformations and diagnosis of Ankylosispondylitis, Primary osteo-arthritis and kyphosis. A patient is asked to bend forward by sliding the hand down to the knee at different positions. They are asked to slide down the side, or rotate the shoulder while pelvis remaining at unchanged position. For all positions using a flexible meter, measurements are performed and clinical estimation is reached.

Here, for the first time known in literature, we simulated the same positions and for each case moiré measurement is performed. The main experimental element to perform MT-based measurement is a diffraction grating. We have made and applied diffraction gratings and performed image processing while applying coherent techniques, we have introduced before [18, 19, 20, 21].We have applied gratings with changeable transparency of their periodic structure. Using phase and changeable amplitude gratings, we have reached to shift the optical transmitted energy towards the higher diffraction orders or towards the higher frequency. At the same time, in the past, we have shown that by using higher diffraction orders, measurement accuracy and sensitivity can be increased. In case of MT application, by using higher diffraction orders fringe multiplications can be performed. In Section 3 of this paper, we have performed MT-based measurement by using a diffraction grating with changeable transparency.

In section 2, the concept regarding moiré application is presented. The theory has been written down in the literature frequently, as it is a well established theory. The present work is exclusively about the experimental work, where Moiré fringes of human body have been performed and analyzed using fringe interpretation technique, where quick-fringe software has been used. Experimental results obtained by using projection moiré are presented in Section 3, where typical cases of 3-D deformations are selected. Subsequently, the main results obtained for the human body at quantitatively determined position using two weighing scales and measurement of the distance of tip of the finger from the knee are presented at this section as well. The fringe interpretation using "quick Fringe " 1-D, 2-D and 3-D information.

The use of Projection Moiré technique

The Moiré figure can be obtained by superposition of two periodic structures. The periodic structures can be superposed to form a new pattern, with new periodicities in addition to those of the individual grids. In Figure 2 the schematic representation of the projection moiré technique is shown.

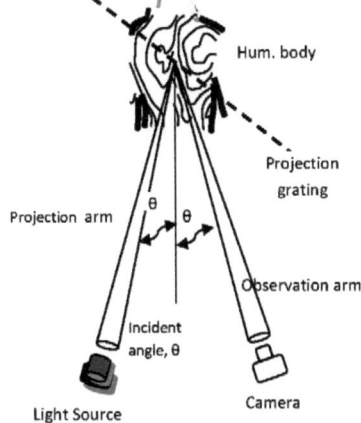

Fig. 2. Setup for projection moire

The projection arm with the light source and the observation arm with camera and the object are shown. The object here is a human back.

The newly obtained sub patterns shown in this figure are the moiré lines. Coarse-structure components in these patterns can be extracted by low-pass filtering or visual observation. Information about the coarse moiré patterns of Fourier transform is contained in the low-frequency components of the Fourier transform of the complex structure. This transmission function is the product of the transmittance functions of the individual grids, and the transform of a product of functions is the convolution of their individual transforms.

Regarding moiré theory, there are many books and mathematically instructive papers [22, 23] that have explained mathematical concept of Moire phenomenon M(x,y) and its application by using three steps:

a) Encoding the image to be studied I(x, y) into a distorted grating J(x,y),

b) Placing a second (reference) grating C(x,y) on top of the (object) image,

c) observing and interpreting the moiré pattern represented by

$$M(x,y)=J(x,y)*(Cx,y)$$

The individual structures are characterized by three parameters: spatial frequency, orientation, and element profile (variable of the transmittance over one period). Moire' can be defined as the convolution or operation in which position vectors of the functions to be convolved are added vectorially, whereas the corresponding profile-shape transforms are multiplied. The transform of the single element of the pattern is usually called the diffraction function; the transform of its periodicity is called the interference function.

EXPERIMENTAL INTERPRETATIONS

Formation of the Moiré on a human back

Initially we have projected a grid on a human back. The body of a young man was placed in a position shown in Fig. 3. As is shown the vertical lines of the projected grid have different lengths according to the regions of the back.

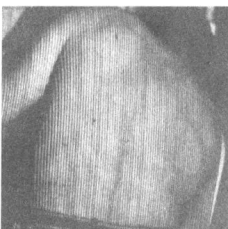

Fig. 3 Projection of a grid on the human back.

The grating used for projection is named as an object image. Using a source of light, the object grating is projected on the surface of a human back. As shown in Fig. 3 the image of the back can be considered as anew grating. The total structure of this grating is actually the same as the shape of the body. Therefore, any change induced on the back and defined as a deformation, could be measured by measuring the change of the newly obtained grating as shown in Fig3.

If the size of the body is increased, the distance between the lines of the projected grid is increased as well. Deformation induced on the body due to bending or twisting for a particular angle will cause change of shape of the grating as shown in Fig. 3. Therefore, the distance between the lines, shape or constant of the projected grating is changed due to the deformation of the human back. The changed parameters of this grating that contain the information of the human back are visualized and quantified or measured, using moiré phenomenon.

In case of Projection Moiré, we have used the setup shown on Fig. 2. The illumination angle is smaller than 90^0. A source of light is placed on one side, let us say on the left side of the line perpendicular to the surface of the body or object. A grating named as object grating is placed between the source and the body. On the other side of the normal to the surface of the object, at certain distance, is placed a camera. Between the surface of the body and camera, a grating named as a reference grating is placed. The superposition information of the grating projected on the surface and this grating named as the reference grating, generates Moiré figures that are photographed by the camera. We have performed Moiré by using the grating with a constant transparency and with changeable transparency. The results are shown in Fig, 4 and Fig. 5 respectively.

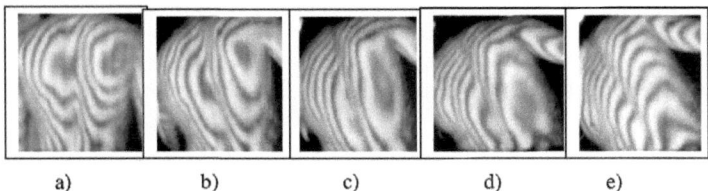

a)　　　　b)　　　　c)　　　　d)　　　　e)

Fig. 4. Moire results obtained when conventional gratings are used..

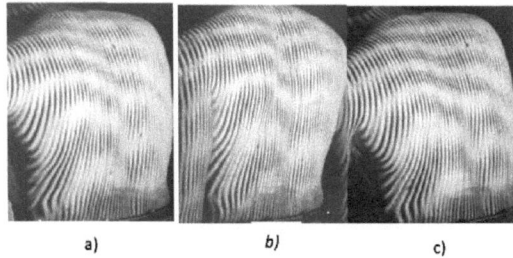

a)　　　　b)　　　　c)

Fig. 5 Projected moiré: when liner transparency of the grating is applied.

The case shown in Fig.5 will be studied in the future work. Here we study only the case when only conventional gratings are used.

A human is placed all the time at the same place and the source of light, grating in between the source and the body, the grating in front of camera and a camera are fixed all the time at the same place, while the body has changed positions by keeping the legs at the same position, in order to introduce deformations. To make this metrology technique more robust, the camera is not placed perpendicular to the plane of view.

Quantitative fringe analysis

In order to quantify and reach the reproducibility of the experiment, a graduate student active in sports is taken as a model for investigation. His back was towards us and his legs were placed on two scales. His weight was 81kg.

To determine the reproducible positions of the body, the student is placed with his two legs on two different scales. To simulate deformations, the weight is increased on the left leg and decreased on the right leg. At the same time the left arm is stretched vertically towards the left knee. Each subsequent phase of the human body, simulating the deformations, is imaged and measured by Moiré approach and quantified by fringe interpretation software.

For the fringe interpretation, for each position representing a particular deformation of the back, the same circle size and selection of the same part of the image of the human back is performed. On each image the same software and the same parameters are applied. Results of the fringe interpretation are presented in three different images shown beside the image of the back with moiré fringes and the selected part of the back with the mentioned circle. The so called fringe centers represented by crosses within the circle are placed automatically by the software, ensuring the reproducibility of the same procedure for different images, representing deformations of the back. Next three images beside the image of human back and the circle containing the information only of the part of the body selected within the circle are shown. The first image, on the left top corner of this part of the back represents the contour representation obtained by the software. Second image shown on the right down corner represents one dimensional profile of the surface and fourth image on the right down corner of the main image, represents the three dimensional surface or deformation of the body. The fifth image is the so

46

called fringe synthesis which can be used as additional information for the correct interpretation of the fringes and actual deformation of the back..

In Fig. 6 the human (a student) of weight 81kg is placed with two legs on two different scales. His weight on the left scale was 45.5 kg and on the left scale was 35.5 kg. On the left leg the weight is for 10 kg more than on the right leg. For this case, the moiré fringes shown on the left corner of Fig 6.a, are not symmetric because the weight of the body including the trunk and back are not distributed symmetrically on both scales.

After the software for fringe interpretation is applied, contour, line profile and 3-D surface plot is represented in the Fig 6.b, 6.c, and 6.d respectively. Synthesized fringe representation is shown in the middle of the lowest row, Fig. 6.e.

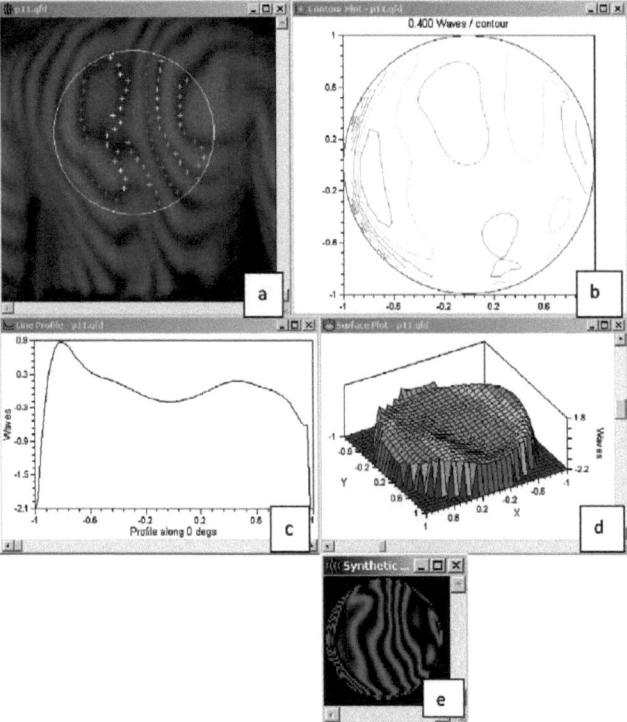

Fig. 6. The case when the weight on the left leg is 45.5 kg and on the right leg is 35.5 kg.

Contour representation obtained by the fringe interpretation software contains a low density of contours indicating that deformations are not emphasized in this case. This can be seen in the surface representation and one-dimensional so called wave representation. Two maximum points related to corresponding points on the human body selected by the circle can be seen. Otherwise, this graph is relatively flat compared to the next cases.

In the next experiment, the weight on the left leg that is measured by the scale is 55 Kg. The position of the body for this case and the moiré fringes are shown at the upper left corner in Fig.7.a. After application of the software for fringe interpretation the contour, line, 3-D surface and synthetic representation are obtained and shown in the Fig. 7.b, Fig.7.c, Fig.7.d, and Fig.7.e, respectively.

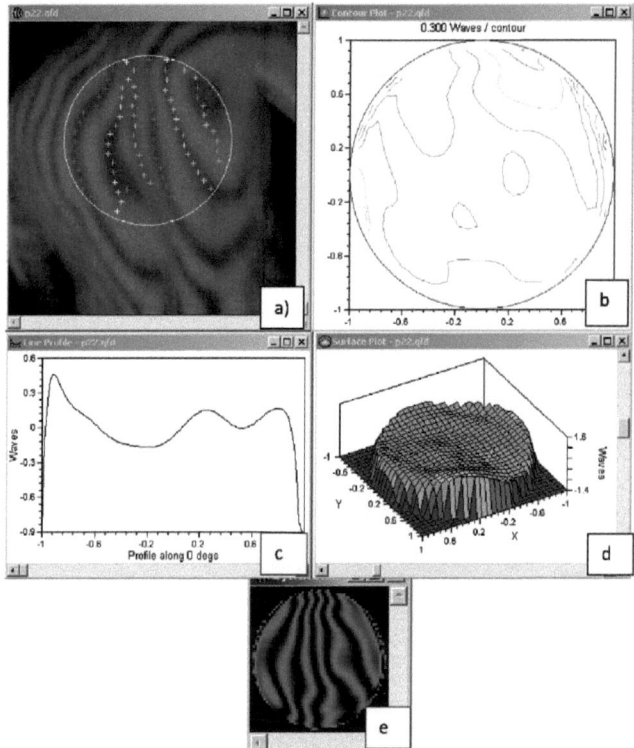

Fig. 7. The case when the weight in the left leg was 55 kg

When the weight is increased on the left leg that is measured 55 kg and the tip of the longest finger of the left hand is only 2 cm above the knee of the left leg. For this position more bending of the body has been required.

In Fig.7. a, is shown position of the body for this case. The moiré fringes obtained for this case are shown as well. After the fringe interpretation is performed by the software, the contour, one dimensional line of the wave representation,2-D surface representation and synthetic representation are shown at the same figure 7. Here the density of the contours is increased, indicating an emphasized deformation of the back. Surface representation and 1-D wave show one maximum and one minimum beside the edges of the contour of the selected part for the human back.

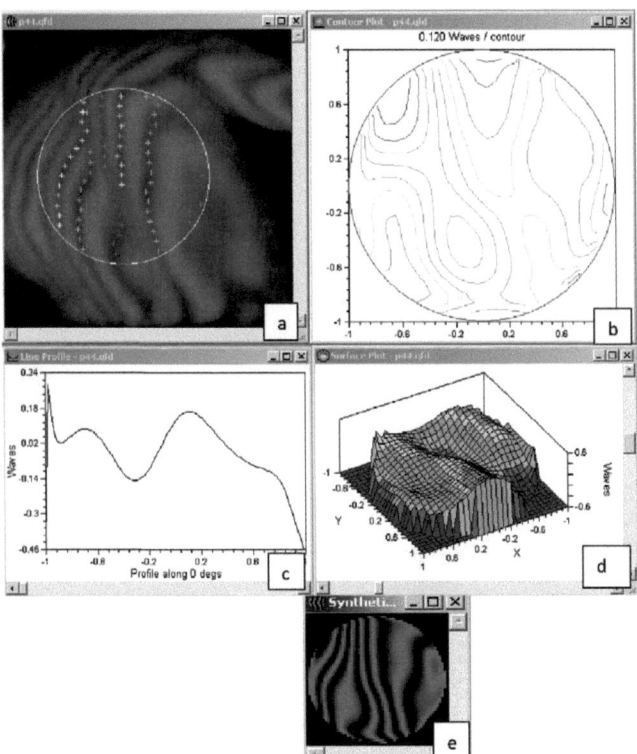

Fig. 8. The case when the weight on the left leg is 68 kg.

In case represented in Fig 8, the weight on the left leg was increased up to 68 kg and the tip of the finger was on the top of the knee. As can be seen on Fig. 8 the shape of the contour representation is changed, indicating a more emphasized deformation of the body. In surface representation and particularly in the image of 1-D wave representation is shown so that two maxima and two minima are obtained. An image of an asymmetric distribution of the fringes is obtained when the fringes are synthesized.

Finally, the total weight of the body is placed on the left leg and is measured 81 kg. At the same time, the longest tip of the finger of the left hand is placed 3 cm below the knee. The position of the body for this case at the upper left corner is shown in Fig.9. Obviously, in-plane and out-of plane deformation has been induced. Deformations are visualized by fringe contours that are obtained on the surface of the body. Analysis of the fringes is performed by using fringe interpretation software. Results for contour, line, 3-D surface and synthetic representation at the same figure are shown.

Very emphasized deformation can be seen from the contour representation of this case as shown in Fig 9. An emphasized minimum can be seen from the 1-D or wave representation of this figure.

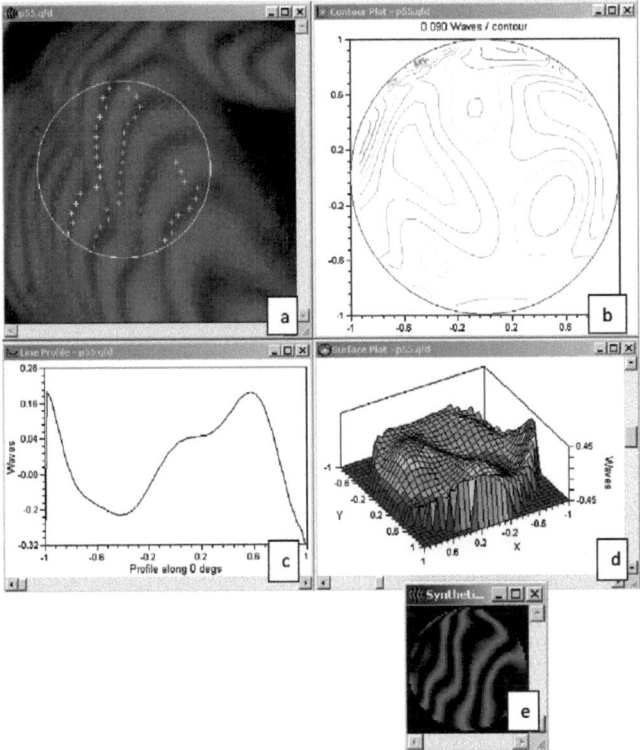

Fig 9. The case when the weight on the left leg is 81 kg.

CONCLUSION

Clinical investigation of a human back usually is followed by x-ray imaging. This procedure of medical checkup gives us only qualitative information. Moreover, x-ray radiation could cause serious health problems. Moire' Technique is a successful non destructive technique, applied for biomechanical application, where electromagnetic waves of longer and safe wavelengths are used.

Projection moiré approach, proposed here, by using "Quick fringe" for fringe interpretation has some advantages compared to other techniques, when other softwares are used. Results obtained herewith contour 1D, 2-D and 3-D representations, give us information for in-plane and out-of-plane deformations.

The main role for successful application of MT plays the grating used and image processing performed. Based on our experience we had in the past, we made holographic, computer controlled and conventional gratings in order to choose the optimal experimental result. We have applied projection MT with conventional and holographic gratings with constant transparency or constant amplitude-based gratings. Moiré outputs are obtained and fringe interpretation is performed, while using those gratings. However, we used our computer generated-gratings (CBG) with changeable transparency, so called amplitude and phase gratings. Results of moiré are shown, when CBG are used. These are attractive, because it can be used to change the direction of the diffracted light of higher energy towards higher diffraction orders of higher frequencies. We indicate that the use of the selected higher diffraction orders can be used for multiplication of the moiré fringes. The increase of measurement sensitivity and accuracy by using CBG will be our task of the future work.

REFERENCES

[1] **Matula P, Kozubek M, Staier F&Hausmann M, 2009.** Precise 3D image alignment in micro-axial tomography. Journal of Microscopy 2003, (Pt 2):126-42

[2] **Hrazdira L&Skotáková J, 2006.** 3D ultrasonography for examination of the musculoskeletal system.Acta chirurgiae orthopaedicae et traumatologiae Cechoslovaca **73**(6):414-20.

[3] **Liu CM & Chen LW, 2004.** Digital atomic force microscope moiré method. Ultramicroscopy**101**(2-4):173-81.

[4] **Rousset MM, Simonek F & Dubus JP**, Radiographic images: appearance and reality (a method of measurement of the 3D coordinates of anatomic points), Bulletin Du Grupment International Pour La Research Scientifiue En Stamatologie Et Odontologie 2000 May-Dec;42(2-3):65-71.

[5] **Sebastian Witt & Das Wunderwerek Mensch**, GEO compact, Nr. 26, 33 (2011).

[6] **Gomes PF, Sesselmann M, Faria CD, Araújo PA, &Teixeira-Salmela LF, 2010.** Measurement of scapular kinematics with the moiré fringe projection technique.Journal of Biomechanics **43**(6):1215-9.

[7] **Falcão AX &Udupa JK. 2000.** A 3D generalization of user-steered live- wire Segmentation,Medical Image Analysis **4**(4):389-402.

[8] **Moran AJ, &Lipczynski R**, Automatic digitization and analysis of moiré topograms on a personal computer for clinical use.Medical Engineeering Physics 1994 May,16(3):259-64.

[9] **Takasaki, H.** Applied Optics Vol. 9, Issue 6, pp. 1467-1472 (1970).

[10] **Takasaki, H.** Applied Optics Vol. 12, Issue 4, pp. 845-850 (1973).

[11] **Flávia Porto, Jonas L. &Gurgel, 2012.** Thaís Russomano and Paulo T.V. Farinatti.Moiré Topography: From Takasaki Till Present Day,*Published9.05.2012,ww.intechopen.com/download/pdf/36611.*

[12] **El-Sayyad M, 1986.**Comparison of roentgenography and moiré topography for quantifying spinal curvature. Physics and Thermodynamics **66**: 1078-1082.

[13] **Batouche M,Benllamr R & Kholland Mk.** **1996,** A computer vision system for diagnosing scoliosis using moiré images. Computers in Biology and Medicine**26** (4): 339-353

[14] **Kim H & Rodrigez M, 2009**. Automatic detection of spinal deformity based on statistical features from the moiré topographic images. Computing,; 8 (1): 72-78.

[15] **Roland McRae, 1978**. Clinical Orthopaedic examination, Hong Kong Printing Press, ISBN 0443015120.

[16] **Richard W. Porter**, Management of the Back Pain, Churchill Livingstone, House, ISBN 0443029547 (1986).

[17] **Kokaj J. and Li Young Quing,1989,** Deformation measurement using FT Technique, Expperimental Technich der Physik, Gemany., **37**, 23-29.

[18] **Kokaj J. Makdisi Y & Bhatia K, 1997** Dental Deformation measurement using holographic interferometry, Optik,No.**1,** 11-16.

[19] **Kokaj J., Makdisi Y & Bhatia K,1997**. Deformation measurement using Moments, Optik, 101 ,**9,**49-52.

[20] **Kokaj J.,2000.** Higher-moment and higher diffraction-order computation for deformation measurement, Optics and Lasers in Engineering, **33**, 165-175

[21] **Kokaj J., 1978.**Application of phase and amplitude gratings in holography and deformation measurement. Theses for Candidate of Science, MGU, Moscow.

[22] **Lohman A. & Sinzinger S.,1993**. Moire affect as a tool for image processing, Journal of Optical Society of America A, vol **10**, No. 1 65-68.

[23] **Theocaris P. S.**, Moiré Fringes in Strain Analysis (Pergamon Press, London, (1969).

*Appendix

Moiré Phenomenon and its application [1, 2, 3]

In this appendix is presented the mathematical explanation of Moire phenomenon and is not written by author. It is not his intellectual property. Here is included in appendix in order to be clear to the reader the concept of Moiré phenomenon. Similar explanations exist in different books and reports. We have selected the explanation provided by Theocaris [1] ,other sources [2].and a NASA report [3].

More detailed information can be seen in [3] and elsewhere . Particular information and application are written: "An instrument development program aimed at using Projection Moiré Interferometry (PMI) for acquiring model deformation measurements in large wind tunnels was begun at NASA Langley Research Center in 1996. Various improvements to the initial prototype PMI systems have been made throughout this development effort. This paper documents several of the most significant improvements to the optical hardware and image-processing software, and addresses system implementation issues for large wind tunnel applications. The improvements have increased both measurement accuracy and instrument efficiency, promoting the routine use of PMI for model deformation measurements in production wind tunnel tests."

Shadow–Moiré

A grating lying over the curved surface is illuminated under the angle of incidence θ_1 (measured from the grating normal) and viewed under an angle θ_2, as shown in Fig 1. From this we can see that a point P_0 on the grating is projected to a point P_1 on the surface which by viewing is projected to the point P_2 on the grating. This is equivalent to a displacement of the grating relative to its shadow equal to a displacement of the grating to its shadow equal to

$$u = u_1 + u_2 \qquad (1)$$

Modulation function $\Psi(x)$ is u/p, where p is the spacing distance in the grating constant

$$\Psi(x) = u/p = \Delta z/p \ (\tan\theta_1 + \tan\theta_2) ,$$

Δz is the height difference between the grating and the point P_0 on the surface.

$$\Delta z = \frac{np}{\tan\theta_1 + \tan\theta_2} \qquad (2)$$

A bright fringe is modulated whenever $\Psi(x) = n$, for $n = 0, 1, 2, \ldots$.

Dark fringes are formed on the surface for $\Delta z = (n+1/2)p \ /(\tan\theta_1 + \tan\theta_2)$

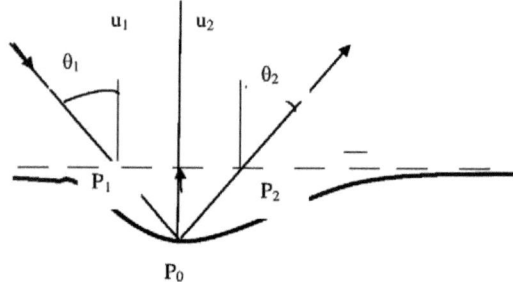

Fig. 2

The result of such a configuration is that the object appears to have not the grid shadows, but contour lanes with the surface of the object. The contour planes are literally a separate spatial frequency generated by beating the spatial frequency of the illumination grid with that of the observation grid. In this case both grids are one and the same, and the contours are planes that are parallel to the plane of the grid (the x-y plane) and have a spacing of p.

Therefore p is the grid spacing, and θ_1 and θ_2 are the illumination and observation angles as illustrated in Fig. 2. If the contour spacing (Δx) is much greater than the grid spacing (p), then the moire contours are easily separable from the grid shadows.

This form of basic moiré contouring is extremely powerful under the right circumstances. The grid must be located relatively close to the object to prevent deleterious diffraction effects. Obviously, placing a grid close to the human body is not a easy task. And while the grid spacing can be widened, which will allow it to be removed from the immediate vicinity of the model, this necessitates an increase in the contour spacing, and, therefore, a reduction in measurement accuracy. Even then, the grid, as well as window through which its shadow is cast, must be as large as the model if full coverage is desired. It is these difficulties that necessitate a modification of the basic moiré technique.

Projection Moiré

The optical configuration of a projection moiré system is illustrated in Fig. 3. It consists mainly of a projection and an observation arm which respectively project and view moiré shadow fringes on the model. The model is centered on the origin of an x,y,z Cartesian coordinate system. Note that the x axis is perpendicular to the plane of the figure. Both the projection and observation arms have optic axes that lie in the y, z plane; intersect the origin; and are removed from one another by an angle 2θ. Note that the z axis bisects this angle, and that there is no loss of generality here since unequal projection and observation angles can be represented by a simple rotational transformation of the model coordinate space. The system operates in the following fashion:

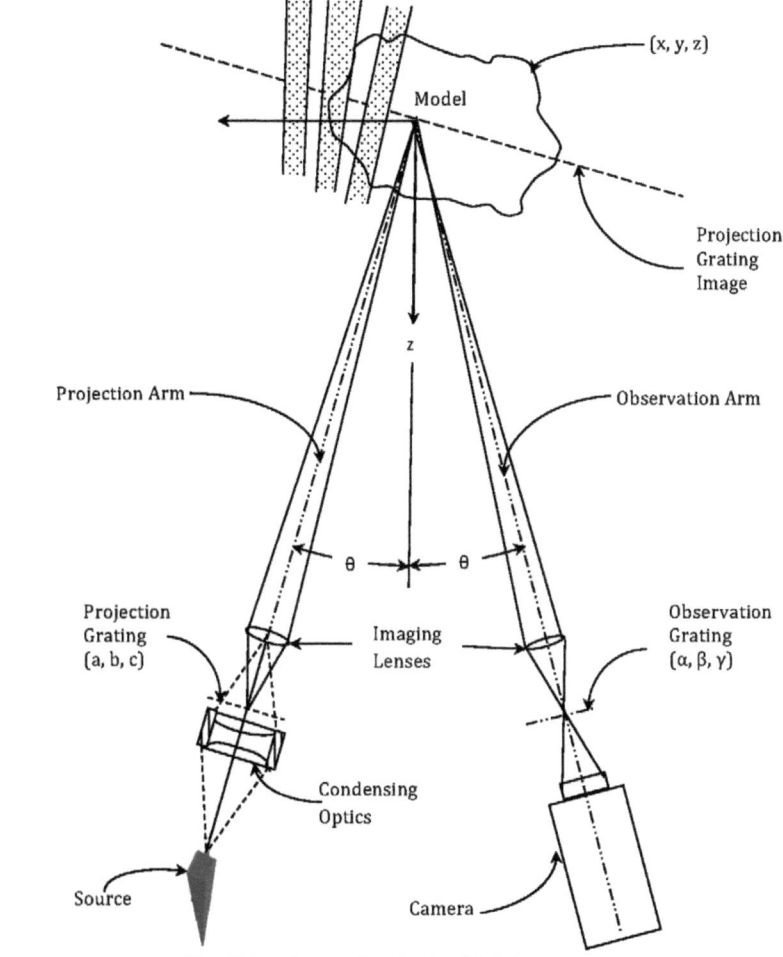

Fig. 3 Experiment of projection Moiré

An incoherent, white light source illuminates the grating in the projection arm, which is in turn imaged onto the model surface. Now while the grating itself is two dimensional, and has no significant longitudinal depth, its image has a longitudinal depth that is equal the depth of focus of the imaging optics, which is chosen to be equal to the maximum depth of the model. Hence, the image of the projection grating in the vicinity of the model consists of dark and light planes which can be, but are not necessarily, parallel to one another. These planes are illustrated by the shaded bands in the model space of Fig.3. Note that for clarity of the illustration, only some of the grating lines have been shaded in this fashion.

If the transmission function of the projection grating is defined as $T_p(a, b, c)$, where a, b, c is the Cartesian coordinate space of the projection grating, and if it is illuminated by the uniform intensity I_s, then the intensity seen by the imaging lens is

$$I_p(a, b, c) = I_s\, T_p(a, b, c) \tag{3}$$

The intensity distribution projected into the model coordinate space is

$$I_p(x, y, z) = I_s\, T_p(x, y, z) \tag{4}$$

Where
$$I_p(a, b, c) \rightrightarrows I_p(x, y, z) \tag{5}$$

represents the imaging transformation from the projection grating to the model coordinate space. At the model, the intensity of the light scattered to the observer is

$$R(x, y, z) = S\,(x, y, z)\, I_p(x, y, z) \tag{6}$$

where $S(x, y, z)$ is the localized point scattering function of the surface of the model, assumed to be diffuse and Lambertain.

Now the imaging lens of the observation arm collects and images this scattered light onto the observation grating. The intensity distribution of this image before passing through the observation grating is $R(\alpha, \beta, \gamma)$ where α, β, γ is the Cartesian coordinate space of the observation grating and

$$R(x, y, z) \rightrightarrows R(\alpha, \beta, \gamma) \tag{7}$$

represents the imaging transformation for the model t o the grating coordinate space. If the intensity transmission function of the observation grating is defined as $T_o(\alpha, \beta, \gamma)$, then intensity distribution seen by the observer is

$$I_o(\alpha, \beta, \gamma) = T_O(\alpha, \beta, \gamma)R(\alpha, \beta, \gamma) \tag{8}$$

And since the imaging transformation of a properly corrected imaging lens is linear, the transformation of Equation 6 is merely,

$$R(\alpha, \beta, \gamma) = S(\alpha, \beta, \gamma) \; I_p(\alpha, \beta, \gamma) \tag{9}$$

Therefore, it follows from Equations 6, 7, 8 and 9 that

$$I_o(\alpha, \beta, \gamma) = I_S \, S(\alpha, \beta, \gamma)T_o(\alpha, \beta, \gamma) \, T_P(\alpha, \beta, \gamma) \tag{10}$$

To interpret Equation 10, which represents the observed image note that $I_s \, S(\alpha, \beta, \gamma)$ is merely the image that would be recorded if both gratings were removed so that

$$T_O = T_P = 1 \tag{11}$$

However, with the gratings in the system, the intensity distribution of the model image is modulated by the combined transmission function

$$T(\alpha, \beta, \gamma) = T_O(\alpha, \beta, \gamma)T_P(\alpha, \beta, \gamma) \tag{12}$$

The moiré fringes are beat spatial frequencies that arise as a result of the multiplication on the right hand side of Equation (12).

To understand fully the manner in which this occurs, it will be necessary to consider some specific grating transmission functions. However, when doing so, it will be more instructive if the observed intensity distribution in Equation (9) is referred to the coordinate space of the model. This merely requires the reverse of the imaging transformation in Equation 7, which is itself an imaging transformation from the observation grating to the model coordinate space. And again, since such a transformation is linear,

$$I_o(x, y, z) = I_S S(x, y, z)T(x, y, z) \qquad (13)$$

Where $\quad T(x, y, z) = T_O(x, y, z)T_P(x, y, z)$

The transformation involved shows that moire phenomenon is equivalent to projecting the shadows of both gratings, onto the surface of the model in such a way that resultant intensity is product of the two projection intensities, rather the sum as would be the actual case. This is the simplest way of understanding and analyzing the moiré phenomenon.

Differential Projection Moiré Contouring:

The objective of differential moiré contouring is to isolate motion of the model surface from its shape. To do so the system of Figure 3 is used with a slight but significant modification. Namely, the simple line grating in the observation arm is replaced with a more complex grating that is representative of the model surface shape, and under the right circumstances the contour fringes that result are a function of model deflection alone.

To fabricate the observation grating for differential moiré, the simple observation grating for standard moiré is removed and replaced with unexposed film. The simple line grating is retained in the projection arm so that the film records a transparency of the projection grating fringes as they appear cast upon the model surface, but perturbed by the surface shape, position, and the angle of view of the observation arm. After processing, this transparency replaces the old observation grating in the otherwise unchanged system of Figure 3. The projection grating, still unchanged, projects fringes onto the model surface, which are again perturbed by the surface shape, position, and the angle of view of the observation arm. If the model surface is unchanged since the differential observation grating was recorded, then the observed perturbations match those in the differential observation grating through which they are viewed, and the difference spatial frequencies cancel. Hence, any difference frequencies present are a measure of change in the model surface shape and position.

60

The fringes, as mentioned are projected onto the model surface by the projection arm. During recording of the differential observation grating the model surface is described by the topographic function $z_0(x, y)$, and the optical intensity distribution at the film plane is (akin to equation 6-10)

$$R_o (\alpha, \beta, \gamma) = I_S S_O (\alpha, \beta, \gamma)T_{PO} (\alpha, \beta, \gamma) \qquad (14)$$

where R, S and T_p are determined after the fashion of Equations 1-13, and

$$R(x, y, z_O) \rightrightarrows R_O (\alpha, \beta, \gamma) \qquad (15)$$

represents the imaging transformation from the model coordinate space to that of the grating. Assuming and ideal film characteristic, the resultant transmission function of the developed film is

$$T_o (\alpha, \beta, \gamma) = R_O (\alpha, \beta, \gamma) \qquad (16)$$

As a later time, with the developed film acting as the differential observation grating, the fringes cast by the projection arm are unchanged, but the model surface is assumed to have moved and/or warped to the new topographic function $z(x, y)$. As before the intensity distribution scattered by the model is $R(x, y, z)$ which, after the imaging transformed becomes $R(\alpha, \beta, \gamma)$. This intensity distribution is filtered by the transmission characteristic.

$$I_o (\alpha, \beta, \gamma) = R (\alpha, \beta, \gamma)R_O (\alpha, \beta, \gamma) \qquad (17)$$

As before, it will be more instructive to relate the observed intensity distribution to the coordinate space of the model by reversing the imaging transformation. The result is

$$I_O (x, y, z) = I_O{}^2 S(x, y, z_O) S(x, y, z)T_P (x, y, z_O)T_P (x, y, z) \quad (18)$$

where z and z_0 are both x and y dependent topographic functions. Using above driven equations on can obtain for the transparence:

$$T(x, y, z) = \frac{1}{4}\left\{1+\cos\left(\frac{2\pi}{P'} L_P \right)+\cos\left(\frac{2\pi}{P'} L_O \right)+1/2\cos\left[\frac{2\pi}{P'}(L_P +L_O)\right]\right.$$

$$+1/2 \cos\left[\frac{2\pi}{P'}(L_P - L_O)\right]\bigg\}$$

(19)

Where, for $L_O(y, z) = L_P(y, z_O)$

One can determine the fringes by

$$L_P = N_P'$$
$$L_O = N_P'$$
$$L_P + L_O = N_P'$$
$$L_P - L_O = N_P'$$

After a proper mathematical calculations on can obtain:

$$\frac{y\cos\theta - z\sin\theta}{d - z\cos\theta - y\sin\theta} \pm \frac{y\cos\theta - z_O\sin\theta}{d - z_O\cos\theta - y\sin\theta} = \frac{N_P'}{d}$$

(20)

where d_P has been replaced by d. Of interest here are the difference spatial frequencies which, are determined by

$$(z - z_O)(y - d\sin\theta) =$$

$$\frac{N_P'}{d}[(d - y\sin\theta)^2 - (d - y\sin\theta)(z + z_O)\cos\theta + zz_O\cos^2\theta]$$

(21)

The terms in the last equation are grouped according to their z and z_O dependence. Note that the left side is a pure function of the difference between the two topographic surfaces $z(x, y)$ and z_O (x, y), whereas the right side is a function of their sum as well as their product. The contour fringes that result, therefore, are a complex function of both surface shape (z and z_O) as well as deflection $(z - z_O)$. Only in the telecentric case (the limit as d approaches ∞) is the shape dependence removed, in which case Equation (21) reduce to

$$(z - z_O)\sin\theta = -N_P'$$

(22)

Like this dependence N_P' of $(z - z_O)$ is reached. The relevant or determining factor is $\sin\theta$, shown in the experimental (Fig. 3).

References

[1] Theocaris P. S., Moiré Fringes in Strain Analysis (Pergamon Press, London, (1969)

[2] Fleming, G., Soto, H., South, B., and Bartram, S., "Advances in Projection Moiré Interferometry Development for Large Wind Tunnel Applications," SAE Technical Paper 1999-01-5598, 1999, doi:10.4271/1999-01-5598.

[3] Gery A. Fleming, Hector L. Soto, Bruce W. South, Scott M. Bartram, NASA Larley Research Center, PMI, (1996)